Agriculture Issues and Policies

Brassicaceae

Characterization, Functional Genomics and Health Benefits

AGRICULTURE ISSUES AND POLICIES

Additional books in this series can be found on Nova's website
under the Series tab.

Additional e-books in this series can be found on Nova's website
under the e-book tab.

AGRICULTURE ISSUES AND POLICIES

BRASSICACEAE

CHARACTERIZATION, FUNCTIONAL GENOMICS AND HEALTH BENEFITS

MINGLIN LANG
EDITOR

New York

For permission to use material from this book please contact us:
Telephone 631-231-7269; Fax 631-231-8175
Web Site: http://www.novapublishers.com

Library of Congress Cataloging-in-Publication Data

ISBN: 978-1-62808-856-4

Published by Nova Science Publishers, Inc. † New York

CONTENTS

Preface

The world now is entering into an ageing society with the number of older people projected to increase, which is predicted to increase over 130% between 2000 and 2050. The particular importance of delivery of health care will thus shift from acute to chronic illnesses. While the high speed of economic development and industrialization make a large area of global soil and water polluted by heavy metals, and it has been a big barrier to the worldwide food production and safety for continuing support the lives of the global increasing population, and high quality of clean living condition requirements. Advances in molecular and cell biology, genetics, genomics and ecology over the last two decades have generated exciting discoveries that consuming and application of plant species from Brassicaceae will solve or prevent most problems addressed above. For example, the Brassicaceous plants derived glucosinolate showed promising effects on preventing chronic diseases such as cancer, cardiovascular and neurodegenerative diseases that affecting mostly older people. Although aspects of Brassica research have been reviewed from time to time, we are not aware of any single book that has covered the breadth and depth of current research in species of Brassicaceae for treating the problem we are facing. The objective of editing this book is to provide the up-to-date references for those interested in and increase the attention on the importance of this Brassicaceae family of crops and plants, and our increasing understanding of the beneficial compounds from the Brassicas and the critical molecular and physiological processes will aid us to breed new varieties to meet the needs of a growing population's health. This book covers 7 chapters which have been well prepared by the leading scientists of the world from China, USA, Argentina, Spain, Portugal and Japan, who have long experience and intensive knowledge of the subjects. This book volume lead us to the frontiers of understanding of the some of the Brassica Functional Genomics and proteomics as they concern critically important structures and functions occurring at the molecular level. We believe, however, that our collaboration on this book volume represents a melding of our perspectives that will provide new dimensions of appreciation and understanding for all researchers and students. I should also like to acknowledge our colleagues Prof. Stefan Hörtensteiner (University of Zürich, Switzerland), Prof. Alan Baker (University of Melbourne, Australia), Prof. Nishanta Rajakaruna (College of The Atlantic, USA) and Dr. María Moreira (Comisión Nacional de Investigaciones Científicas y Técnicas (CONICET), Argentina), who carefully reviewed the selected chapters.

In: Brassicaceae
Editor: Minglin Lang

ISBN: 978-1-62808-856-4

Chapter 1

HEALTH BENEFITS OF *BRASSICA* SPECIES

Tzi Bun Ng,* Charlene Chiu Wing Ng and Jack Ho Wong
School of Biomedical Sciences, Faculty of Medicine, The Chinese University of Hong Kong, Shatin, New Territories, Hong Kong, China

ABSTRACT

Phenethylisothiocyanate produced by *Brassica* food plants is known to produce various health benefits. Oral administration of PEO, a phenethylisothiocyanate essential oil containing more than 95% natural phenethylisothiocyanate, was effective in causing remittance of acute and chronic signs of ulcerative colitis in mice. The varieties of two *Brassica* species, "grelos" (rape) and "espigos" ("tronchuda" cabbage) are nutritionally well-balanced vegetables. "Tronchuda" cabbage has the highest levels of β-carotene, vitamin C, moisture, proteins, and fat. Rape has the highest contents of ash, carbohydrates, chlorophylls, flavonoids, lycopene, phenolics, sugars (including fructose, glucose, sucrose and raffinose), tocopherols, α-linolenic acid, the best ratios of polyunsaturated to saturated fatty acids, and the highest antioxidant properties. Antifungal proteins from seeds of various *Brassica* species including *B. oleracea, B. campestris, B. juncea* var. *integrifolia, B. parachinensis,* and *B. alboglabra* suppressed proliferation of cancer cells. Some of them exerted antifungal activity against the yeast *Candida albicans* and the fungus *Fusarium oxysporum*, exhibited antibacterial activity against *Pseudomonas aeruginosa*, and reduced the activity of HIV-1 reverse transcriptase. Napin-like polypeptides from seeds of *B. chinensis* cv dwarf and *B. alboglabra* exhibited antibacterial activity. Napin-like polypeptide from *B. parachinensis* seeds manifested antiproliferative activity against cancer cells and stimulated nitrite production by mouse peritoneal macrophages. The cruciferous vegetables broccoli, cabbage and cauliflower are abundant in phytochemicals such as glucosinolates and their byproducts, phenolics and antioxidant vitamins and dietary minerals. The organosulfur chemicals namely glucosinolates and the S-methyl cysteine sulphoxide found in broccoli in concert with other constituents such as vitamins E, C, K and the minerals such as iron, zinc, selenium and the polyphenols namely kaempferol, quercetin glucosides and isorhamnetin are presumably responsible for various health benefits of broccoli. The

* Corresponding author. School of Biomedical Sciences, Faculty of Medicine, The Chinese University of Hong Kong, Shatin, New Territories, Hong Kong,China. Email : b021770@mailserv.cuhk.edu.hk.

health benefits associated with their antioxidant properties signify the importance of dietary intake of these vegetables.

Keywords: *Brassica,* vegetables, health benefits, phytochemicals, medicinal plants

INTRODUCTION

Cruciferous vegetables such as broccoli, cauliflower, cabbage, kale, mustard and turnip are popular all over the world. It is well known that a copious intake of vegetables and fruits is beneficial to health and that prevention of diseases is better than cure. The intent of the present article is to review literature pertaining to health promoting constituents of (Brassicaceae) vegetables.

HEALTH BENEFITS OF *BRASSICA* PLANTS IN GENERAL

Among the various subspecies of *Brassica oleracea,* kale had the highest content of the antioxidants carotene, tocopherol, and ascorbate, followed by broccoli and Brussels sprouts with moderate levels, and then by cauliflower and cabbage, with comparatively low concentrations (Kurilich *et al.*, 1999).

There is a correlation between copious intake of cruciferous vegetables and a lowered risk of lung and gastrointestinal cancer. Glucosinolates in cruciferous vegetables and its metabolites, the isothiocyanates and nitriles, modify enzymes regulating xenobiotic metabolism, and induce cell cycle arrest and apoptosis. It is believed that a combination of a variety of cruciferous vegetables may offer optimal protection (Lund, 2003).

Indole-3-carbinol (I3C) and phenethylisothiocyanate (PEITC), hydrolytic products of *Brassica* plants with anti-cancer property, promoted bile excretion and enhanced bileγ-GTP activity in the first 24 h after treatment in rats. This finding is noteworthy since bile has cancer-chemopreventive action (Ishibashi *et al.*, 2012).

PEO is a PEITC Essential Oil containing over 95% natural PEITC. Orally administered PEO was effective at alleviating acute and chronic signs of ulcerative colitis in mice. It improved body weight and stool consistency and reduced, mucosal inflammation, depletion of goblet cells, infiltration of inflammatory cells and intestinal bleeding, as well as production of proinflammatory interleukin-1beta. The disease attenuation by PEO is likely associated with reduction of total cellular Signal Transducer and Activator of Transcription 1 (STAT1) as well as nuclear phosphorylated-STAT1 (activated form of STAT1), decrease of mRNA of C-X-C motif ligand 10 (a STAT1 responsive chemokine) and interleukin 6. PEO might be a promising candidate to develop as a treatment for ulcerative colitis patients (Dey *et al.*, 2010).

HEALTH BENEFITS OF BROCCOLI (*BRASSICA OLERACEA VAR. ITALICA*)

The vegetables broccoli, cauliflower and cabbage are abundant in the phytochemicals glucosinolates and their byproducts, phenolics, antioxidant vitamins, and dietary minerals.

Consumption of broccoli will provide antioxidants, regulate enzymes and regulate apoptosis and cell cycle. Glucosinolates and the S-methyl cysteine sulphoxide in broccoli, together with other components such as vitamins E, C, K, the minerals selenium, zinc, iron, and the polyphenols isorhamnetin, kaempferol, and quercetin glucosides account for the various health benefits of broccoli (Vasanthi *et al.*, 2009). Results of epidemiological studies suggest that consumption of cruciferous vegetables like cabbages and broccoli leads to a diminished cancer risk due to the presence of specific glucosinolates, a group of sulphur-containing glucosides (Heaney and Fenwick, 1995).

Epidemiological evidence discloses health benefits resulting from the consumption of broccoli, especially with regard to chemoprevention. Since broccoli is abundant in selenium and glucosinolates (especially glucoraphanin and isothiocyanate sulforaphane), which produce the redox-regulated cardioprotective protein thioredoxin (Trx), broccoli consumption may trigger cardioprotection. Cardioprotection after broccoli consumption is indicated in the ischemic/reperfused rat heart by better post-ischemic ventricular function, suppressed cardiomyocyte apoptosis, reduced cytochrome c release, elevated pro-caspase 3 activity, and smaller myocardial infarct size. RNA transcripts and protein levels of the thioredoxin superfamily comprising Trx1,Trx2, glutaredoxin Grx1, Grx2, and peroxiredoxin (Prdx), were either reinstated or augmented following broccoli consumption. Broccoli enhanced the expression of Nrf2, a cytosolic Keap1 suppressor, indicating the involvement of antioxidant response element in Trx induction. Broccoli upregulated the expression of heme oxygenase-1, a cardioprotective protein that is transactivated during Trx activation. Broccoli brought about Akt phosphorylation and Bcl2 induction together with activation of redox-sensitive transcription factor NFkappa B and Src kinase, suggesting the participation of Akt, Bcl2, and cSrc in generating the survival signal (Mukherjee *et al.*, 2008).

Mikkelsen *et al.* (2012) have developed a platform for stable expression of multi-gene pathways in *Saccharomyces cerevisiae*. Introduction of the seven-step pathway of indolylglucosinolate from *Arabidopsis thaliana* to the yeast, *S. cerevisiae* enabled the first successful microbial production of glucosinolates. Large-scale production for the benefit of human health thus appears to be feasible.

Broccoli (*Brassica oleracea* var. *italica*) accumulates high levels of Ses. Se-methylselenocysteine which is one of the most effective chemopreventive compounds as the predominant selenoamino acid. A cDNA encoding selenocysteine Se-methyltransferase, the key enzyme contributing to SeMSC formation, was cloned from broccoli using an *Arabidopsis thaliana* homocysteine S-methyltransferase gene probe, and the clone (BoSMT) was functionally expressed in *Escherichia coli*. The BoSMT transcript and SeMSC synthesis were low in level in selenite-treated plants but up-regulated in selenate-treated plants. Treatment of selenate with selenite undermined SeMSC formation. Elevated levels of sulfate suppressed selenate uptake, with a consequent marked decline in BoSMT mRNA level and SeMSC accumulation. SeMSC accumulation closely correlated with BoSMT gene expression. The total Se status in tissues provides important information for maximizing the SeMSC production in broccoli (Lyi *et al.*, 2005).

Selenium (Se)-fortified broccoli has been promoted as a functional food, which means food to which health promoting substances have been added. After exposure of plants to 20 μM sodium selenate, nearly 50% of total Se in the foliage was due to Se-methylselenocysteine and selenomethionine. Glucosinolate content remained unaltered. Essential micronutrients comprising Cu, Fe, Mn, and Zn were unchanged among 50% of the germplasm. Total antioxidant capacity was substantially enhanced in over half of the accessions. Thus breeding of broccoli cultivars that accumulate Se and other compounds beneficial to health is possible (Ramos *et al.*, 2011).

HEALTH BENEFITS OF KALE (*BRASSICA OLERACEA* L. VAR. *ACEPHALA*)

Selenium (Se) is a micronutrient in mammalian nutrition and is accumulated in kale (*Brassica oleracea* L. var. *acephala*), which has high levels of lutein and beta-carotene. Selenium, beta-carotene and lutein are powerful antioxidants and have health benefits (Lefsrud et al., 2006). Increases in either selenate or selenite resulted in decreases in kale leaf tissue biomass. Neither selenate nor selenite treatment affected lutein or beta-carotene accumulation in leaves. Increasing selenate promoted the accumulation of kale leaf Se; however, leaf tissue Se did not significantly change after the selenite treatments. Increases in selenate affected the leaf tissue concentrations of P, K, Ca, Mg, S, B, Cu, Mn, and Mo, whereas selenite only affected B and S. Growing kale in the presence of selenate would bring about the accumulation of high tissue Se levels without any effect on carotenoid concentrations (Lefsrud et *al.*, 2006).

Brassica oleracea var. *acephala* has been employed in Brazilian traditional medicine for treating gastric ulcer (Carvalho *et al.*, 2011, Lemos *et al.*, 2011). A hydroalcoholic extract of its leaves did not exert genotoxic or clastogenic effects on murine brain cells, bone marrow cells hepatocytes, leukocytes, and testicular cells. However, it was capable of mitigating doxorubicin-induced DNA damage. The antigenotoxic activity of this extract may have some value for cancer prevention (Gonçalves *et al.*, 2012).

HEALTH BENEFITS OF RED CABBAGE (*BRASSICA OLERACEA* VAR. *CAPITATA*)

Green varieties of cabbage (*Brassica oleracea* var. *capitata*) have little, if any, anthocyanin. Red cabbage is red due to the presence of anthocyanin which demonstrates a positive correlation with total antioxidant power (Yuan *et al.*, 2009).

Polyphenol extracts from *Brassica* vegetables (Brussels sprouts and red cabbage) lowered cholesterol concentrations and extent of lipid peroxidation in hypercholesterolemic erythrocytes but not in control normal erythrocytes. Membrane fluidity remained unaltered after treatment in both normal and hypercholesterolemic erythrocytes (Duchnowicz *et al.*, 2012).

Sulforaphane (SFN), an isothiocyanate formed by hydrolysis of glucosinolates found in *Brassica oleraceae*, is reported to possess anticancer and antioxidant activities. SFN isolated from red cabbage *(Brassica oleraceae* var. *rubra*) down-regulated the expression of bcl-2 (antiapoptotic), while up-regulating p53 and Bax (proapoptotic) proteins cells in HEp-2 human epithelial carcinoma cell line (Devi and Thangam, 2012).

HEALTH BENEFITS OF WHITE CABBAGE (*BRASSICA OLERACEA* VAR. *CAPITATA* CV. TALER)

The content of glucosinolates, ascorbigen, and ascorbic acid in white cabbage (*Brassica oleracea* var. *capitata* cv. Taler) varied depending on the season (summer or winter), fermentation, and salt concentration used for brining (0.5% NaCl or 1.5% NaCl). Different salt concentrations were used for sauerkraut salt concentration production. Glucobrassicin, glucoiberin, and sinigrin were found to be dominant in raw white cabbage cultivated either in winter or in summer. The content of the ascorbigen precursor glucobrassicin was about 40% higher in winter cabbage than summer cabbage. Cabbage fermented for 7 d had very little glucosinolates regardless of the fermentation conditions used. A low salt concentration (0.5% NaCl) raised ascorbigen content in sauerkraut after fermentation at 25° C for one week. The highest ascorbigen concentration was noted in low-sodium (0.5% NaCl) sauerkraut produced from winter cabbage submitted to either natural (109.0 micromol/100 g distilled water) or starter-induced fermentation (108.3 and 104.6 micromol/100 g distilled water) in cabbages fermented by *Lactobacillus plantarum* and *Leuconostoc mesenteroides*, respectively). Ascorbic acid content was found higher in summer cabbage and reduced by fermentation. Hence, cabbages with high glucobrassicin content and low-sodium sauerkrauts may be beneficial to health (Martinez-Villaluenga *et al.*, 2009).

HEALTH BENEFITS OF RAPE AND "TRONCHUDA" CABBAGE

The varieties of two *Brassica* species, known in Northern Portuguese regions as “grelos” (rape) and “espigos” (“tronchuda” cabbage) are vegetables with a good nutritional value. “Tronchuda” cabbage exhibited the highest levels of proteins, β-carotene, vitamin C, fat, and moisture. Rape had the largest amounts of ash, chlorophylls, flavonoids, lycopene, phenolics, tocopherols, carbohydrates, sugars (including fructose, glucose, sucrose and raffinose), the essential n-3 fatty acid α-linolenic acid, and the best ratios of polyunsaturated to saturated fatty acids and n-6/n-3 fatty acids, as well as the highest antioxidant activity (Batista *et al.*, 2011).

Rapeseed oil phenolics, principally vinylsyringol, effectively scavenged radicals and inhibited production of proinflammatory prostaglandin E(2). There was no mutagenicity or toxicity to Caco-2 cells or macrophages (Vuorela *et al.*, 2005).

Plant sterols and their hydrogenated forms, stanols, can bring about a reduction in serum low density lipoprotein-cholesterol levels. In *Brassica* species, brassicasterol is the predominant sterol. *Streptomyces hygroscopicus* 3-hydroxysteroid oxidase has been utilized to engineer rapeseed (*Brassica napus*) oilseeds to change the relative amounts of specific sterols to stanols. The major phytosterols were reduced at the C-5 double bond to the corresponding phytostanol without affecting the C-22 double bond (Venkatramesh *et al.*, 2003).

Nanotechnology which produces particles such as liposomes and nanoliposomes made of pure phospholipids is used in pharmaceutics to augment drug bioavailability and bioefficiency. Rapeseed lecithin liposomes improve cell proliferation in rat bone marrow stem cells (Arab Tehrany *et al.*, 2012).

HEALTH BENEFITS OF INDIAN MUSTARD (*BRASSICA JUNCEA*)

Very long chain polyunsaturated fatty acids (VLCPUFAs) such as (AA), (EPA) and docosahexaenoic acid (DHA) are valuable commodities that provide important human health benefits. Wu *et al.* (2005) reported the transgenic production of significant amounts of arachidonic acid AA and eicosapentaenoic acid EPA in *Brassica juncea* seeds via a stepwise metabolic engineering strategy.

Vitamin A deficiency has led to an elevated risk of severe morbidity and mortality in some countries such as India (Chow *et al.*, 2010) and sub-Saharan Africa (Sablah *et al.*, 2012). Consumption of oil from genetically modified mustard (*Brassica juncea*) overexpressing the vitamin A precursor beta-carotene would be in line with WHO recommendations of periodic, high-dose vitamin A supplementation to prevent vitamin A deficiency (Chow *et al.*, 2010).

Mustard oil massage of newborns is a component of traditional care practices in many communities (Darmstadt and Saha, 2003). However, this practice may produce adverse effects, especially in preterm infants and in those with sub-optimal skin barrier function. Other natural oils such as sunflower, sesame or safflower seed oil may have a beneficial effect on neonate health and survival. Mullany *et al.* (2005) administered a questionnaire on the use and rationale for applying mustard oil and other oils to neonatal skin to the caretakers of 8580 neonates in Sarlahi district of rural Nepal. It was found that about 99% neonates received mustard oil massage at least once in the first two postnatal weeks, and 80% received two or more massages daily. Mustard oil was applied for promoting strength, maintaining health, and giving warmth. An understanding of cultural, social, and economic factors that shape the context of traditional healthcare practices is essential to the design and implementation of intervention trials examining the relative efficacy of application of oils in reducing neonatal mortality and morbidity (Mullany *et al.*, 2005).

Black mustard is used as a spice and an inexpensive source of antimicrobial agents for treating bacterial infections (Dubie *et al.,* 2012). Rajamurugan *et al.* (2012) reported that the crude methanol extract of black mustard (*B. nigra*) leaf is nontoxic and protects against the toxicity of d-galactosamine on the rat kidneys and liver as indicated by reduction in serum levels of urea, uric acid, creatinine, and bilirubin levels, and tissue levels of thiobarbutric acid reactive substance, enzymic and non-enzymic antioxidants and inflammatory marker enzymes such as myeloperoxidase, cathepsin D, and acid phosphatase. Decrease in hepatic and renal damage is observed in histopathological studies.

HEALTH BENEFITS OF CANOLA (*BRASSICA NAPUS*)

Zhang *et al.* (2007) studied how the total concentration and the composition of tocopherols and phytosterols in canola seedlings and extracted oil were affected by seed germination under illuminated and dark environments.

A net increase in alpha-tocopherol and total tocopherols indicating new tocopherol synthesis was observed from day 10 to day 20 of germination under illumination. However, in the dark no net increase in tocopherol was noted. Tocopherols were concentrated in the leafy seedling apex and not in the non-photosynthetic base, unlike phytosterols which were equally distributed. The total tocopherol content of oil extracted from 20-day-old seedlings was 4.3- to 6.5-fold higher than that of intact seeds over the sprouting period, but the concentration of total phytosterols in the oil fraction increased 4.2- to 5.2-fold. The concentration of these valuable phytochemicals in the oil fraction is attributed mainly to the exhaustion of oil reserves that occurs during germination, and the light-induced de novo alpha-tocopherol synthesis. Thus germination is a way to naturally concentrate these high-value constituents in canola oil (Zhang *et al.*, 2007).

Investigations on canola seeds overexpressing the bacterial phytoene synthase gene (crtB) have shown a 50-fold rise in the total carotenoid level, comprising phytoene and downstream metabolites like beta-carotene, with a 2:1 beta- to alpha-carotene ratio. There was a 90% decline in phytoene levels for the double construct expressing phytoene synthase (crtB) and phytoene desaturase (crtI). Transgenic seeds from all double constructs, including that expressing the bacterial crtB and the plant lycopene beta-cyclase, exhibited augmented levels of total carotenoid analogous to that previously noticed by expressing crtB alone but little effects were detected with regard to the beta- to alpha-carotene ratio in comparison with the original construct. However, the ratio rose from 2:1 to 3:1 when a triple construct encompassing bacterial phytoene synthase, phytoene desaturase and lycopene cyclase genes were co-expressed. The data indicate that the bacterial genes may form an aggregate complex which permits *in vivo* activity of all three proteins through substrate channeling. Thus further manipulation of the carotenoid biosynthetic pathway may lead to downstream products with elevated agronomic, animal feed and human nutritional values (Ravanello *et al.*, 2003).

Brassica Antifungal Proteins

A 9412-Da antifungal lipid transfer protein from *Brassica campestris* seeds inhibited mycelial growth in *Mycosphaerella arachidicola* and *Fusarium oxysporum* with an IC(50) value of 4.5 microM and 8.3 microM, respectively (Lin *et al.*, 2007). Another 9.4-kDa thermostable and pH-stable antifungal lipid transfer peptide designated as campesin exerted an inhibitory action on mycelial growth including *F. oxysporum* and *M. arachidicola*, with an IC(50) of 5.1 microM and 4.4 microM, respectively. It inhibited the activity of HIV-1 reverse transcriptase with an IC(50) of 3.2 microM, and proliferation of HepG2 and MCF cancer cells with an IC(50) of 6.4 microM and 1.8 microM (Lin *et al.*, 2009). Lin *et al.* (2007) compared LTP isolated from *B. campestris* seeds with mung bean LTP and chitinase. The antifungal activity of *Brassica* and mung bean LTPs were thermostable, pH-stable, and unaltered following exposure to proteases. The antifungal activity of mung bean chitinase was much less pH- and thermo- stable. *Brassica* LTP but neither mung bean LTP nor mung bean chitinase inhibited proliferation of hepatoma Hep G2 cells and breast cancer MCF 7 cells and the activity of HIV-1 reverse transcriptase.

A 5907-Da thermostable and pH-stable antifungal peptide from kale (*Brassica alboglabra*) seeds inhibited mycelial growth in fungi *Valsa mali, Helminthosporium maydis, Mycosphaerella arachidicola* and *Fusarium oxysporum*, with an IC(50) of 0.15 microM, 2.1 microM, 2.4 microM, and 4.3 microM, respectively. It inhibited the activity of HIV-1 reverse transcriptase with an IC(50) of 4.9microM and proliferation of breast cancer (MCF7) and hepatoma (HepG2) cells with an IC(50) of 3.4 microM and 2.7 microM, respectively (Lin and Ng, 2008).

An 5716 Da thermostable and pH-stable antifungal peptide from *Brassica parachinensis* designated as brassiparin potently inhibited mycelial growth in *Fusarium oxysporum, Helminthosporium maydis, Mycosphaerella arachidicola* and *Valsa mali*. It inhibited proliferation of hepatoma (HepG2) and breast cancer (MCF7) cells and the activity of HIV-1 reverse transcriptase (Lin and Ng, 2009).

An 18.9 kDa antifungal protein designated as juncin from Japanese takana (*Brassica juncea* var. integrifolia) seeds exhibited antifungal activity toward the phytopathogens *Mycosphaerella arachidicola, Fusarium oxysporum,* and *Helminthosporium maydis,* with IC(50) values of 10,13.5, and 27μM, respectively. It inhibited the activity of HIV-1 reverse transcriptase with an IC(50) of 4.5 μM , and the proliferation of hepatoma (HepG2) and breast cancer (MCF7) cells with IC(50) values of 5.6 and 6.4 μM, respectively (Ye and Ng, 2009).

A 30 kDa protein purified from red cabbage (*Brassica oleracea*) seeds hindered mycelial growth in *Mycosphaerella arachidicola* (with an IC50=5 μM), *Setospaeria turcica,* and *Bipolaris maydis*. It also inhibited the yeast *Candida albicans* with an IC50=96 μM. It exerted its antifungal action by permeabilizing the fungal membrane as evidenced by staining with Sytox green. The antifungal activity was stable from pH 3 to 11 and from 0 to 65 °C. It manifested antibacterial activity against *Pseudomonas aeruginosa* (IC50=53 μM). Furthermore, after 48 h of culture, it suppressed proliferation of nasopharyngeal cancer and hepatoma cells with IC50=50 and 90 μM, respectively (Ye *et al*., 2011).

Brassica Napin-Like Polypeptides

Napins are 1:1 disulfide-linked complexes of a smaller subunit and a larger subunit. A heterodimeric 13.8 kDa napin-like polypeptide from Chinese cabbage (*Brassica parachinensis*) seeds manifested higher trypsin inhibitory than chymotrypsin inhibitory activity. It stimulated nitrite production by murine peritoneal macrophages and reduced the viability of leukaemia (L1210) cells (Ngai and Ng, 2004a). The polypeptide potently exhibited cell-free translation-inhibiting activity in a system with an IC50 of 6.2 nM. The polypeptide was relatively stable in the pH range 6-11 and in the temperature range 10-50 degrees C (Ngai and Ng 2003).

A heterodimeric napin-like polypeptide from kale seeds exhibited antibacterial activity against *Bacillus, Megabacterium,* and *Pseudomonas* species and antiproliferative activity against leukemia L1210 cells. It inhibited translation in the rabbit reticulocyte lysate system with an IC50 of 37.5 nM. This activity was retained between pH 5 and pH 11, and between 10 and 40°C, but declined to low levels at pH 3 and pH 13 and at 70° C (Ngai and Ng, 2004b).

A heterodimeric 11-kDa napin-like polypeptide from Chinese white cabbage (*Brassica chinensis* cv dwarf) seeds manifested antibacterial activity against *Pseudomonas aeruginosia, Bacillus subtilis, Bacillus cereus,* and *Bacillus megaterium.* It inhibited translation in the rabbit reticulocyte system with an IC50 of 18.5nM.This translation-inhibitory activity was stable between pH 4 and 11, and between 10 and 40°C. The polypeptide inhibited trypsin with a higher potency (IC50 = 8.5 microM) than it inhibited chymotrypsin (IC50 = 220 microM) (Ngai and Ng, 2004c).

Cultivation of Broccoli

Domínguez-Perles *et al.* (2010) carried out an investigation on biologically active compounds (glucosinolates, phenolic acids, and flavonoids), nutrients (vitamin C, minerals, and trace elements), and *in vitro* radical-scavenging capacity of harvest remains obtained from greenhouse cultivation of broccoli. The cultivation was conducted using 80 mM NaCl treatment, typical of the irrigation water in the production areas of Murcia located in the Southeast part of Spain. The bioactive compounds and nutrient contents varied depending on the cultivar, organ (foliage or stalks), and the saline stress (80 mM NaCl), in three different cultivars Marathon, Nubia, and Viola. Cultivar Nubia was not affected by 80 mM NaCl treatment to any marked extent. The phytochemical and nutrient contents in the cultivation byproducts of Nubia were similar to health-promoting levels of edible commercial parts (inflorescences or flower heads). Agrowaste recycling to yield biologically active ingredients for industry can raise profit, cut cost and minimize environmental problems (Domínguez-Perles *et al.*, 2010).

Conversion of Cauliflower Byproducts into High-added Value Compounds

Green labeled pectins were extracted by using proteases and cellulases to digest cellulose and proteins in the cell wall. High methoxy and low methoxy pectins of high molar mass isolated from cauliflower florets and leaves were demethylated with *Aspergillus aculeatus* pectin methyl esterase. Health benefit pectic oligosaccharides were obtained after enzymatic treatment of the residue recovered after pectin extraction. The enzymatic method indicates the fesibility of converting vegetable byproducts into high-added value compounds, such as pectins and pectic oligosaccharides, and thus considerably reduce the quantity of these residues produced by food industries (Zykwinska *et al.*, 2008).

Interspecies and Stage-dependent Variation of Content Health-promoting Compounds of *Brassica* Species

Park *et al.* (2012) observed that the amounts of glucosinolates, anthocyanins, carotenoids, and other secondary metabolites in the skin and flesh of pale green and purple kohlrabi (*Brassica oleracea* var. *gongylodes*) varied greatly between the two types of kohlrabi.

Sasaki *et al.* (2012) employed a C30 column and an ammonium formate buffer in LC-MS and a micro plate solid phase extraction technique to determine the levels of glucoraphanin which is a precursor of sulforaphane, an isothiocyanate well known for its potential health benefits. The glucoraphanin level found in three cabbage cultivars and six kale cultivars were similar to, or even higher than, the highest of broccoli (119.4 mg/100g fresh weight).

Antioxidant activity of six *Brassica* crops including broccoli, cabbage, cauliflower, kale, nabicol and tronchuda cabbage was the highest at three months after sowing. Kale crop exhibited maximal antioxidant activity also at the adult stage. The peak antioxidant activity in cauliflower also occurred in sprouts and in leaves taken two months after sowing. Variation in antioxidant activity of *Brassica* crops were associated with differences in total phenolic content and also to differences in phenolic composition. Brassica by-products could be utilized as sources of products with high antioxidant activity (Soengas *et al.* 2012).

Total and individual glucosinolate (GSL) content of the leaves of turnip rape (*Brassica rapa* L. var. *rapa*) was measured in 45 varieties comprising early, medium and late types cultivated at two locations in northwestern Spain. Two most abundant GSLs were gluconapin and glucobrassicanapin which account for 84.4 % and 7.2 % of the total GSL content, respectively. The highest total GSL content was found in the varieties, MBG-BRS0429 and MBG-BRS0550 (from turnip greens and extra-late groups) and MBG-BRS0438 (from turnips and late groups). Breeding strategies should be designed for producing GSL-rich varieties (Cartea *et al.* 2012).

Potential Health Risks of Broccoli

Excessive intake of glucosinolates may impede growth, impair performance and affect renal, hepatic, and thyroidal function in pigs but not in humans (Heaney and Fenwick, 1995).

Isothiocyanates and indoles in broccoli are glucosinolate-derived degradative products that arise as a consequence of the catalytic action of plant myrosinase and/or glucosidases derived from the human microbial flora. Besides anticarcinogenic activity, these products might also have adverse effects, especially genotoxic activities. Latté *et al.* (2011) gave an overview on genotoxic, anti-genotoxic, chemopreventive, nutritive and antinutritive properties of broccoli, its ingredients and their degradation products. It appears that modest intake is beneficial. Regarding diets with exceptionally high daily intake, fortified broccoli-based dietary supplements, and raw consumption of broccoli, the potential risks and beneficial effects await assessment.

Epithiospecifier Protein from Broccoli (*Brassica oleracea* L. ssp. *italica*) Inhibits Formation of the Anticancer Agent Sulforaphane

Sulphoraphane, a major isothiocyanate in broccoli seedlings, potently induces phase 2 detoxification enzymes. However, epithiospecifier proteins (non-catalytic cofactors of myrosinase) may also favor the generation of the non-inductive sulphoraphane nitrile (Williams et al., 2008).

In broccoli (*Brassica oleracea* L. ssp. *italica*), in the presence of epithiospecifier protein (ESP), epithionitrile is formed as a result of myrosinase -catalyzed hydrolysis of alkenyl glucosinolates such as sulforaphane. Epithionitrile production is negatively correlated with formation of the sulforaphane. A 43-kDa protein with ESP activity and manifesting sequence homology to *Arabidopsis thaliana* ESP was cloned from broccoli cv. Packman and expressed in *Escherichia coli*. It directed myrosinase-dependent metabolism of the alkenyl glucosinolate epi-progoitrin [(2S)-2-hydroxy-3-butenyl glucosinolate] to form an epithionitrile as well as myrosinase-dependent hydrolysis of the glucosinolate glucoraphanin [4-(methylsulfinyl)butyl glucosinolate] to form sulforaphane nitrile, instead of isothiocyanate sulforaphane. Sulforaphane but not sulforaphane nitrile has anticarcinogenic properties. Genetic manipulation designed to attenuate or eliminate expression of ESP in broccoli could increase the conversion of glucoraphanin to sulforaphane, enhancing potential health benefits (Matusheski *et al.*, 2006).

There is a requirement to accurately determine the levels of glucoraphanin in vegetable products. Broccoli seeds, which have an abundance of glucosinolates, particularly glucoraphanin, are good for the isolation of glucoraphanin. A novel preparative scale HPLC method with simple compound recovery has been developed to meet the need for a glucoraphanin standard (Rochfort *et al.*, 2005).

EFFECT OF PROCESSING ON HEALTH-PROMOTING COMPOUNDS OF *BRASSICA* SPECIES

In broccoli an increment of sulforaphane content as well as antioxidant activity is noted following steaming and drying, most likely due to an increase of the extractability of antioxidants and sulforaphane. On the other hand, polyphenol concentration is diminished after freezing and boiling, largely owing to volatilization and leaching into the cooking water. Thus broccoli processing should be optimized to maximize the content of bioactive compounds (Mahn and Reyes, 2012).

Roasting of high erucic mustard (HEM) seeds before oil extraction confers a special flavor and augments the oxidative stability of the extracted oil. Compared with rapeseeds, HEM varieties (*Brassica juncea* , *B. juncea* var. *oriental*, *B. nigra* , and *Sinapis alba*) produce during roasting less than 33% of canolol (2,6-dimethoxy-4-vinylphenol which is a powerful radical scavenging compound), owing to a reduced free sinapic acid content and a diminished loss of sinapic acid derivatives. Approximately half of the canolol produced in the roasted seed is extracted into the oil. Thus roasting of HEM seeds can be employed to produce canolol-enriched oil (Shrestha *et al.*, 2012).

Processing brought about, in both green and red cultivars of curly kale (*Brassica oleracea* L. convariety. *acephala* variety. *sabellica*), a decline of total phenolics, antioxidant capacity, and content and distribution of flavonols, anthocyanins, hydroxycinnamic acids, glucosinolates,and vitamin C. In contrast, the red curly kale cultivar was better able to withstand heat-induced destruction of phytochemicals. The extracts of both green and red curly kale inhibited the cell proliferation of Caco-2, HT-29, and HCT 116 human colon cancer cells. Extracts from fresh plant material was more potent in antiproliferative activity than extracts from processed plant material (Olsen *et al.*, 2012).

The Winterbor F(1) variety of kale (*Brassica oleracea* L. var. *acephala*) has good nutritive value and high antioxidant activity. Cooking reduced the antioxidant activity especially the activity of vitamin C and polyphenols and to a smaller extent β-carotene. Hence it is advisable to eat the vegetable in the raw form or have minimal processing prior to eating (Sikora and Bodziarczyk, 2012).

Crucifers contain very high concentrations of glucosinolates (β-thioglucoside-N-hydroxysulfates). Although not themselves protective, glucosinolates GS are converted by coexisting myrosinases to bitter isothiocyanates (ITC) which defend plants against predators (Fahey *et al.*, 2001).. Coincidentally, ITC also induce mammalian genes that regulate defenses against oxidative stress, inflammation, and DNA-damaging electrophiles (Hecht, 2000; Brown and Hampton, 2011; Mi *et al.*, 2011).Consequently, the efficiency of conversion of GS to ITC may be critical in controlling the health-promoting benefits of crucifers. If myrosinase is heat-inactivated by cooking, the gastrointestinal microflora converts GS to ITC, a process abolished by enteric antibiotics and bowel cleansing (Fahey *et al.*, 2012).

Table 1. Summary of health promoting actions of Brassicaceae plants

Plant name	Activities	References
Broccoli	Cardioprotective	(Mukherjee *et al.*, 2008)
	Anticancer	(Matusheski *et al.,* 2006)
	Antioxidant	(Kurilich *et al.*, 1999)
Kale	Antigenotoxic	(Gonçalves et al., 2012)
	Anti-ulcer	(Carvalho et al., 2011, Lemos et al., 2011)
	Antioxidant	(Kurilich *et al.*,1999)
	Antifungal	(Lin and Ng, 2008)
	Antibacterial	(Ngai and Ng, 2004b)
	Anticancer	(Ngai and Ng, 2004b)
Red cabbage	Lowers cholesterol content and reduces lipid peroxidation in erythrocytes exposed to a high cholesterol environment	(Duchnowicz *et al.*, 2012)
	Anticancer	(Devi and Thangam, 2012)
	Antioxidant	(Yuan *et al.*, 2009)
	Antifungal	(Ye *et al.*, 2011).
	Antiproloferative	(Ye *et al.*, 2011).
White cabbage	Antioxidant	(Martinez-Villaluenga *et al.*, 2009)
Chinese cabbage	Immunostimulatory	(Ngai and Ng, 2004a)
Canola	Antioxidant	(Zhang *et al.*, 2007)
Rape	Antioxidant	(Vuorela *et al.*,2005; Batista *et al.*, 2011),
	Lowers serum low density lipoprotein-cholesterol level	(Vuorela *et al.*,2005)
Black mustard	Renprotective, hepatoprotective and antioxidant	(Rajamurugan *et al.* , 2012)
	Antimicrobial	(Dubie *et al.,* 2012)
Japanese tanaka	Antifungal, antiproloferative	(Ye and Ng, 2009),

Broccoli extracts exposed to microwave for 0, 1, and 4 min possessed 9.5, 25.5, and 0 micromol/L sulforaphane (sulphoraphane, a major isothiocyanate in broccoli seedlings, potently induces phase 2 detoxification enzymes) and induced greater than two-fold changes in expression of 381, 1017, and 101 genes in Caco-2 cells, respectively. Seventy-two genes comprising genes regulating polyamine catabolism and transforming growth factor-beta signaling displayed analogous alterations in expression after treatment with all 3 extracts. The concentrations of putrescine and N-acetyl-spermine were upregulated, and the TGFbeta1-mediated induction of phosphorylated Smad 2 was inhibited (Furniss *et al.*, 2008).

Conclusion

Brassica vegetables contain a variety of phytochemicals that have health promoting effects including flavonols, anthocyanins, β-carotene, hydroxycinnamic acids, glucosinolates, and vitamin C. Glucosinolates are converted by coexisting myrosinases to bitter isothiocyanates which induce mammalian genes that regulate defenses against oxidative stress, inflammation, and DNA-damaging electrophiles. The health promoting effects comprise protection against oxidative damage as well as chemical-induced hepatic and renal damage, and reduction of risk of lung and gastrointestinal cancer. Substances with health promoting effects have been purified from *Brassica* species including antifungal proteins, napin-like polypeptides, glucosinolates and small molecules such as flavonols, anthocyanins, β-carotene, hydroxycinnamic acids, and vitamin C.

It is known that processing will result in a loss of the health promoting phytochemicals in *Brassica* vegetables. A copious intake of these plants will undoubtedly bring health benefits. More health benefits in addition to those listed in Table 1 will certainly be disclosed as research continues.

References

Ahmed, A.S., Saha. S.K., Chowdhury, M.A., Law, P.A., Black, R.E., Santosham, M, .Darmstadt.G.L., (2007). Acceptability of massage with skin barrier-enhancing emollients in young neonates in Bangladesh. *J. Health Popul. Nutr.* 25, 236-240.

Arab Tehrany, E., Kahn, C.J., Baravian, C., Maherani, B., Belhaj, N., Wang, X., Linder, M., (2012). Elaboration and characterization of nanoliposome made of soya; rapeseed and salmon lecithins: application to cell culture. *Colloids Surf B Biointerfaces* 95,75-81.

Batista, C., Barros, L., Carvalho, A.M., Ferreira, I.C., (2011). Nutritional and nutraceutical potential of rape (*Brassica napus* L. var. napus) and "tronchuda" cabbage (*Brassica oleraceae* L. var. costata) inflorescences. *Food Chem. Toxicol.* 49, 1208-1214.

Brown, K.K., Hampton, M.B., (2011). Biological targets of isothiocyanates. *Biochim. Biophys. Acta* 1810, 888-894.

de Campos, S.B., Youn, J.W., Farina, R., Jaenicke, S., Jünemann, S., Szczepanowski, R., Beneduzi, A., Vargas, L.K., Goesmann, A., Wendisch, V.F., Passaglia, L.M., (2013).Changes in root bacterial communities associated to two different development stages of canola (*Brassica napus* L. var oleifera) evaluated through next-generation sequencing technology. *Microb Ecol.* 65,593-601.

Cartea, M.E., de Haro, A., Obregón, S., Soengas, P., Velasco, P., (2012). Glucosinolate variation in leaves of *Brassica rapa* crops. *Plant Foods Hum. Nutr.* 67, 283-288.

Carvalho, C.A., Fernandes, K.M., Matta ,S.L., Silva, M.B., Oliveira, L.L., Fonseca, C.C., (2011). Evaluation of antiulcerogenic activity of aqueous extract of *Brassica oleracea* var. capitata (cabbage) on Wistar rat gastric ulceration. *Arq Gastroenterol*. 48, 276-282.

Charron, C.S., Clevidence, B.A., Britz, S.J., Novotny, J.A., (2007). Effect of dose size on bioavailability of acylated and nonacylated anthocyanins from red cabbage (*Brassica oleracea* L. var. capitata). *J. Agric. Food Chem.* 55, 5354-5362.

Chelpanova, T.I., Vitiazev, F.V., Mikhaleva, NIa., Efimtseva, É.A., (2012). Effect of pectin substances on activity of human pancreatic alpha-amylase *in vitro. Ross FiziolZh Im. I. M. Sechenova*. 98,734-743.

Chow, J., Klein, E.Y., Laxminarayan, R., (2010). Cost-effectiveness of "golden mustard" for treating vitamin A deficiency in India. *PLoS One.*;5:e12046.

Darmstadt, G.L., Saha, S.K., (2003). Neonatal oil massage. *Indian Pediatr.* 40.1098-1099.

Devi, J.R., Thangam, E.B., (2012). Mechanisms of anticancer activity of sulforaphane from *Brassica oleracea* in HEp-2 human epithelial carcinoma cell line. *Asian Pac. J. Cancer Prev.* 13, 2095-2100.

Dey, M., Kuhn, P., Ribnicky, D., Premkumar, V., Reuhl, K., Raskin, I., (2010). Dietary phenethylisothiocyanate attenuates bowel inflammation in mice. *BMC Chem. Biol*. 10:4.

Domínguez-Perles, Martínez-Ballesta, M.C., Carvajal, M., García-Viguera, C., Moreno, D.A. (2010). Broccoli-derived by-products--a promising source of bioactive ingredients. *J. Food Sci.* 75, C383-C392

Dubie, J., Stancik, A., Morra, M., Nindo, C., (2013). Antioxidant extraction from mustard (*Brassica juncea*) seed meal using high-Intensity ultrasound. *J. Food Sci.* 78, E542-548.

Duchnowicz, P., Bors, M., Podsędek, A., Koter-Michalak, M., Broncel, M., (2012). Effect of polyphenols extracts from *Brassica* vegetables on erythrocyte membranes (*in vitro* study). *Environ. Toxicol. Pharmacol.* 34,783-790.

Fahey, J.W., Zalcmann, A.T., Talalay, P., (2001). The chemical diversity and distribution of glucosinolates and isothiocyanates among plants. *Phytochemistry* 56,5-51.

Fahey, J.W., Wehage, S.L., Holtzclaw, W.D., Kensler, T.W., Egner, P.A., Shapiro, T.A., Talalay, P., (2012). Protection of humans by plant glucosinolates: efficiency of conversion of glucosinolates to isothiocyanates by the gastrointestinal microflora. *Cancer Prev. Res. (Phila)* 5,603-611.

Finley, J.W., Sigrid-Keck, A., Robbins, R.J., Hintze, K.J., (2005).Selenium enrichment of broccoli: interactions between selenium and secondary plant compounds. *J. Nutr.* 135, 1236-1238.

Furniss, C.S., Bennet, R.N., Bacon, J.R., LeGall, G., Mithen, R.F., (2008). Polyamine metabolism and transforming growth factor-beta signaling are affected in Caco-2 cells by differentially cooked broccoli extracts. *J. Nutr.*138,1840-1845.

Gonçalves, Á.L., Lemos, M., Niero, R., de Andrade, S.F., Maistro, E.L., (2012).Evaluation of the genotoxic and antigenotoxic potential of *Brassica oleracea* L. var. acephala D.C. in different cells of mice. *J. Ethnopharmacol.* 143,740-745.

Hao, X., Mir, P.S., Shah, M.A., Travis, G.R., (2005). Influence of canola and sunflower diet amendments on cattle feedlot manure. *J. Environ. Qual*, 3,1439-1445.

Heaney, R.K., Fenwick, G.R., (1995). Natural toxins and protective factors in brassica species, including rapeseed. *Nat. Toxins* 3, 233-237.

Hecht, S.S., (2000). Inhibition of carcinogenesis by isothiocyanates. *Drug Metab. Rev.*32, 395-411.

Ishibashi, H., Kuwahara, T., Nakayama-Imaohji, H., Ohnish, Y., Mori, H., Shimada, M., (2012). Effects of indole-3-carbinol and phenethyl isothiocyanate on bile and pancreatic juice excretion in rats. *J. Med. Invest.* 59,246-252.

Kurilich, A.C., Tsau, G.J., Brown, A., Howard, L., Klein, B.P., Jeffery, E.H., Kushad, M., Wallig, M.A., Juvik, J.A., (1999). Carotene, tocopherol, and ascorbate contents in subspecies of *Brassica oleracea. J. Agric. Food Chem.* 47,1576-1581.

Latté, K.P., Appel, K.E., Lampen, A., (2011). Health benefits and possible risks of broccoli-an overview. *Food Chem. Toxicol.* 49, 3287-3309.

Lefsrud, M.G., Kopsell, D.A., Kopsell, D.E., Randle, W.M., (2006). Kale carotenoids are unaffected by, whereas biomass production, elemental concentrations, and selenium accumulation respond to, changes in selenium fertility. *J. Agric. Food Chem.* 54,1764-1771.

Lemos ,M., Santin, J.R., Júnior, L.C., Niero, R., Andrade, S.F., (2011). Gastroprotective activity of hydroalcoholic extract obtained from the leaves of *Brassica oleracea* var. acephala DC in different animal models. *J. Ethnopharmacol.* 138, 503-507.

Lin, P., Xia, L., Wong, J.H., Ng,T.B., Ye, X., Wang, S., Shi, X., (2007). Lipid transfer proteins from *Brassica campestris* and mung bean surpass mung bean chitinase in exploitability. *J. Pept. Sci.* 13, 642-648.

Lin, P., Ng, T.B. (2008). A novel and exploitable antifungal peptide from kale (*Brassica alboglabra*) seeds. *Peptides.* 29, 1664-1671.

Lin, P., Ng, T.B., (2009). Brassiparin, an antifungal peptide from *Brassica parachinensis* seeds. *J. Appl. Microbiol.* 106, 554-563.

Lin, P., Wong, J.H., Xi, L., Ng, T.B., (2009). Campesin, a thermostable antifungal peptide with highly potent antipathogenic activities. *J. Biosci. Bioeng.* 108, 59-265.

Lin, P., Xia, L., Ng, T.B., (2007). First isolation of an antifungal lipid transfer peptide from seeds of a *Brassica* species. *Peptides.* 28, 1514-1519.

Lund, E., (2003).Non-nutritive bioactive constituents of plants: dietary sources and health benefits of glucosinolates. *Int. J. Vitam. Nutr. Res.* 73, 135-143.

Lyi, S.M., Heller, L.I., Rutzke, M., Welch, R.M., Kochian, L.V., Li, L., (2005). Molecular and biochemical characterization of the selenocysteine Se-methyltransferase gene and Se-methylselenocysteine synthesis in broccoli. *Plant Physiol.* 138, 409-420.

Mahn, A., Reyes, A., (2012). An overview of health-promoting compounds of broccoli (*Brassica oleracea* var. italica) and the effect of processing. *Food Sci. Technol.* Int. 51803-514.

Martinez-Villaluenga, C., Peñas, E., Frias, J., Ciska, E., Honke, J., Piskula,,M.K., Kozlowska, H., Vidal-Valverde, C., (2009). Influence of fermentation conditions on glucosinolates, ascorbigen, and ascorbic acid content in white cabbage (*Brassica oleracea* var. capitata cv. Taler) cultivated in different seasons. *J. Food Sci.* 74, C62-67.

Matusheski, N.V., Swarup, R., Juvik, J.A., Mithen, R., Bennett, M., Jeffery, E.H., (2006) Epithiospecifier protein from broccoli (*Brassica oleracea* L. ssp. italica) inhibits formation of the anticancer agent sulforaphane. *J. Agric. Food Chem.*54, 2069-2076.

Mi, L., Di Pasqua ,A.J., Chung, F.L., (2011). Proteins as binding targets of isothiocyanates in cancer prevention. *Carcinogenesis*. 32, 1405-1413.

Mikkelsen, M.D., Buron, L.D., Salomonsen, B., Olsen,E., Hansen, B.G., Mortensen, U.H., Halkier B.A., (2012). Microbial production of indolylglucosinolate through engineering of a multi-gene pathway in a versatile yeast expression platform. *Metab. Eng*. 14, 104-111.

Mukherjee, S., Gangopadhyay, H., Das, D.K., (2008). Broccoli: a unique vegetable that protects mammalian hearts through the redox cycling of the thioredoxin superfamily. *J. Agric. Food Chem.* 56, 609-617.

Mullany, L.C., Darmstadt, G.L., Khatry, S.K., Tielsch, J.M., (2005). Traditional massage of newborns in Nepal: implications for trials of improved practice. *J. Trop. Pediatr.* 1, 82-86.

Ngai, P.H., Ng, TB., (2003). Isolation of a napin-like polypeptide with potent translation-inhibitory activity from Chinese cabbage (*Brassica parachinensis* cv green-stalked) seeds. *J. Pept. Sci.* 9, 442-449.

Ng, T.B., Ngai, P.H., (2004a). The trypsin-inhibitory, immunostimulatory and antiproliferative activities of a napin-like polypeptide from Chinese cabbage seeds. *J. Pept. Sci.* 10, 103-108.

Ngai, P.H., Ng, T.B., (2004b) A napin-like polypeptide with translation-inhibitory, trypsin-inhibitory, antiproliferative and antibacterial activities from kale seeds. *J. Pept. Res.* 64, 202-208.

Ngai, P.H., Ng, T.B., (2004c). A napin-like polypeptide from dwarf Chinese white cabbage seeds with translation-inhibitory, trypsin-inhibitory, and antibacterial activities. *Peptides* 25, 171-176.

Olsen, H., Grimmer, S., Aaby, K., Saha, S., Borge, G.I., (2012). Antiproliferative effects of fresh and thermal processed green and red cultivars of curly kale (*Brassica oleracea* L. convar. acephala var. sabellica). *J. Agric Food. Chem*. 60, 7375–7383.

Park, W.T., Kim, J.K., Park, S., Lee, S.W., Li, X., Kim, Y.B., Uddin, M.R., Park, N.I., Kim, S.J., Park, S.U., (2012). Metabolic profiling of glucosinolates, anthocyanins, carotenoids, and other secondary metabolites in kohlrabi (*Brassica oleracea* var. gongylodes). *J. Agric. Food Chem.* 60, 8111-8116.

Rajamurugan, R., Suyavaran, A., Selvaganabathy, N., Ramamurthy, C.H., Reddy, G.P., Sujatha, V., Thirunavukkarasu. C, (2012). *Brassica nigra* plays a remedy role in hepatic and renal damage. *Pharm Biol*. 50, 1488-1497.

Ramos, S.J., Yuan, Y., Faquin, V., Guilherme, L.R., Li, L., (2011). Evaluation of genotypic variation of broccoli (*Brassica oleracea* var. italic) in response to selenium treatment. *J. Agric. Food Chem.* 59, 3657-3665.

Ravanello, M.P., Ke, D., Alvarez, J., Huang, B., Shewmaker, C.K., (2003). Coordinate expression of multiple bacterial carotenoid genes in canola leading to altered carotenoid production. *Metab. Eng*. 5, 255-263.

Robbins, R.J., Keck, A.S., Banuelo, G., Finley, J.W., (2005). Cultivation conditions and selenium fertilization alter the phenolic profile, glucosinolate, and sulforaphane content of broccoli. *J. Med. Food.* 8, 204-214.

Rochfort, S., Caridi, D., Stinton, M., Trenerry, V.C., Jones. R., (2006). The isolation and purification of glucoraphanin from broccoli seeds by solid phase extraction and preparative high performance liquid chromatography. *J. Chromatogr. A.* 1120, 205-210.

Sablah, M., Klopp, J., Steinberg, D., Touaoro, Z., Laillou, A., Baker, S., (2012). Thriving public-private artnership to fortify cooking oil in the West African Economic and Monetary Union (UEMOA) to control vitamin A deficiency: Faire Tache d'Huile en Afrique de l'Ouest. *Food Nutr. Bull.* 33(4 Suppl), S310-320.

Shrestha, K., Stevens, C.V., De Meulenaer, B., (2012). Isolation and identification of a potent radical scavenger (Canolol) from roasted high erucic mustard seed oil from Nepal and its formation during roasting. *J. Agric. Food Chem.* 60, 7506-7512.

Sikora, E., Bodziarczyk, I., (2012). Composition and antioxidant activity of kale (*Brassica oleracea* L. var. acephala) raw and cooked. *Acta Sci. Pol. Technol. Aliment.* 11, 239-248.

Taveira, M., Pereira, D.M., Sousa, C., Ferreres, F., Andrade, P.B., Martins, A., Pereira, J.A., Valentão, P.,.(2009). *In vitro* cultures of *Brassica oleracea* L. var. costata DC: potential plant bioreactor for antioxidant phenolic compounds. *J. Agric. Food Chem.* 57, 1247-1252.

Vasanthi, H.R., Mukherjee, S., Das, D.K., (2009). Potential health benefits of broccoli-a chemico-biological overview. *Mini Rev. Med. Chem* 9,749-759.

Venkatramesh, M., Karunanandaa, B., Sun, B., Gunter, C.A., Boddupalli, S., Kishore, G.M., (2003). Expression of a Streptomyces 3-hydroxysteroid oxidase gene in oilseeds for converting phytosterols to phytostanols. *Phytochemistry* 62, 39-46.

Vuorela, S., Kreander, K., Karonen, M., Nieminen, R., Hämäläinen, M., Galkin, A., Laitinen, L., Salminen, J.P., Moilanen, E., Pihlaja, K., Vuorela, H., Vuorela, P., Heinonen, M., (2005). Preclinical evaluation of rapeseed, raspberry, and pine bark phenolics for health related effects. *J. Agric. Food Chem.* 53, 5922-5931.

Williams, D.J., Critchley, C., Pun, S., Nottingham, S., O'Hare, T.J., (2008). Epithiospecifier protein activity in broccoli: the link between terminal alkenyl glucosinolates and sulphoraphane nitrile. *Phytochemistry* 69, 2765-2773.

Wu, G., Truksa, M., Datla, N., Vrinten, P., Bauer, J/, Zank, T., Cirpus, P., Heinz, E., Qiu, X.. (2005). Stepwise engineering to produce high yields of very long-chain polyunsaturated fatty acids in plants. *Nat. Biotechnol.* 23, 1013-1017.

Ye, X.J., Ng, T.B., Wu, Z.J., Xie, L.H., Fang, E.F., Wong, J.H., Pan, W.L., Wing, S.S., Zhang, Y.B., (2011). Protein from red cabbage (*Brassica oleracea*) seeds with antifungal, antibacterial, and anticancer activities. *J. Agric. Food Chem.* 59, 10232-10238.

Yuan, Y., Chiu, L.W., Li, L., (2009). Transcriptional regulation of anthocyanin biosynthesis in red cabbage. *Planta* 230, 1141-1153.

Zhang, H., Vasanthan, T., Wettasinghe, M., (2007). Enrichment of tocopherols and phytosterols in canola oil during seed germination. *J. Agric. Food Chem.* 55, 355-359.

Zykwinska, A., Boiffard, M.H., Kontkanen, H., Buchert, J., Thibault, J.F., Bonnin, E., (2008). Extraction of green labeled pectins and pectic oligosaccharides from plant byproducts. *J. Agric. FoodChem.* 56, 8926-8935.

Ye, X., Ng, T,B., (2009). Isolation and characterization of juncin, an antifungal protein from seeds of Japanese Takana (*Brassica juncea* var. integrifolia). *J. Agric. Food Chem.* 57, 4366-4371.

In: Brassicaceae
Editor: Minglin Lang

ISBN: 978-1-62808-856-4

Chapter 2

BENEFITS OF *BRASSICA* NUTRACEUTICAL COMPOUNDS ON HUMAN HEALTH

Elsa M. Gonçalves**, *Carla Alegria and Marta Abreu
Instituto Nacional de Investigação Agraria e Veterinária (INIAV), Lisbon, Portugal

ABSTRACT

Due to the many health benefits associated with fruits and vegetables, international dietary recommendations support their increased consumption. People ingest a vast diversity of pharmacologically active chemicals components, nutritional and medicinal, in the form of fruits and vegetables. The consumption of Brassicaceae in particular, contributed in a relevant manner to human nutrition and for other health benefits, as several epidemiological studies have indicated. *Brassica* cruciferous vegetables include different genus of cabbage, cauliflower, collard, broccoli, Brussels sprouts, kale, mustard and rape. All these vegetables supply dietary fiber, and fiber intake is linked to lower incidence of cardiovascular disease and obesity, and also supply vitamins and minerals to the diet. These vegetables are also recognized as sources of phytochemicals that function as antioxidants, phytoestrogens, antiinflammatory agents and other protective compounds associated with a reduced risk of age-related chronic illnesses, such as cardiovascular and other degenerative diseases. In the present review, the predominant members of these biologically active and chemically diverse compounds found in brassicas is addressed. Since the content for these vegetable components varies significantly with plant variety and maturity at harvest, edible organs (*e.g.* roots, shoots, leafs), agriculture practices, postharvest storage conditions and processing methods, the influence of all these factors are reported. Finally we discuss some additional support and strategies to increase and encourage the consumption of brassica vegetables as part of disease risk reduction and healthful eating.

Keywords: *Brassica* vegetables, bioactive compounds, pre- and post-harvest factors, processing, consumption strategies

---

* Corresponding Author address: Instituto Nacional de Investigação Agrária e Veterinária, I.P. Unidade de Investigação de Tecnologia Alimentar. Estrada do Paço do Lumiar, 22. Ed. S, Lisboa. Tel: +351 217127100 fax: +351 217127162. Email: elsa.goncalves@iniav.pt.

Introduction

Brassicaceae family, also known as *Cruciferae*, is a large group having about 3000 species grouped in 350 genera, including several types of edible plants. Petals of plants from this family have a distinctive cross form arrangement, which is the origin of the initial term '*Cruciferae*'. These plants may be annuals, biennials or perennials (Cartea *et al.*, 2011).

The genus *Brassica*, economically speaking, is the most important genus within the tribe Brassiceae as well as considered the most important nutraceutical crops in Europe and America (Wei, Miao, & Wang, 2011). Although essentially temperate, *Brassica oleracea* forms are now grown all over the world (Vaughan & Geissler, 1997). The main species that are commonly grown within *Brassica oleracea*, include vegetable and forage forms, such as kale, cabbage, broccoli, Brussels sprouts, cauliflower and others, while *Brassica rapa* include vegetable forms, such as turnip, Chinese cabbage and pak choi, along with forage and oilseed types. As for *Brassica napus*, these crops are mainly used as oilseed (rapeseed), although forage and vegetable types like leaf rape and 'nabicol' are also included. Finally, the mustard group, which is formed by three species, *Brassica carinata*, *Brassica nigra* and *Brassica juncea*, are mainly used as condiments because of their seeds, although leaves of *Brassica juncea* are also consumed as vegetables (Wei, Miao, & Wang, 2011).

These vegetables are important sources of a variety of nutrients and health-promoting phytochemicals (Liu, 2004) and so have been the focus of numerous epidemiological and clinical studies (Podsedek, 2007) due to their antioxidant and anticarcinogenic properties (Chu *et al.*, 2002; Cohen, Kristal, & Stanford, 2000; Verhoeven *et al.*, 1997). The antioxidant potential of brassica vegetables is high compared to other vegetable crops, containing both hydrophilic and lipophilic antioxidants. Phenolic compounds and ascorbic acid are the major contributors (as hydrophilic antioxidants) due to their high content and high antioxidant activity (Podsedek, 2007). Other vitamins, especially vitamin E (tocopherol) and carotenoids are also important as lipophilic antioxidant compounds in these vegetables (Fahey *et al.*, 2001; Cao *et al.*, 1996). In addition, brassica vegetables provide a large group of glucosinolates, a group of sulphur- and nitrogen-containing secondary metabolites, which have rather low antioxidant activity, but the products of their hydrolysis (namely isothiocyanates) can protect against cancer (Jahangir *et al.*, 2009a; Keum, Jeong, & Kong, 2004; Paolini, 1998, Plumb *et al.*, 1996).

Thus far, some of the more promising anticarcinogenic dietary compounds have been identified in brassica vegetables and further studies of the related protective mechanisms will contribute to support the consumption of these crops (Jahangir *et al.*, 2009a). However, it is necessary to assess the inherent content variation (dependent on pre- and post-harvest conditions, processing, storage or food preparation) of both nutrients and health-promoting phytochemicals in order to better understand the potential health benefits of these crops. In this review the significance brassica vegetables as a source of bioactive compounds for human nutrition and health is made according to their representativeness and function.

Methods

The following review is based on the evaluation of electronically collated data published on brassica phytochemical compounds between 1972 and 2013. It contains 334 references dealing with bioactive compounds and health promoting properties of brassicas, pre-, postharvest and processing effects on its bioactive composition and consumer strategies to improve dietary. Furthermore, data from previous work obtained by the authors on brassica quality are also reported.

1. Bioactive Compounds and Health Promoting Properties of Brassicas

Gathering high quality products with a healthy diet, safety, and convenience is something that consumer's look forward. In addition to the commercial value of the fresh vegetable market, growing interest on the produce bioactive value has risen in growers and processors to specifically reach a health-oriented market.

Over the last two decades, crops in the Brassicaceae have been the focus of intense research based on their health benefits (Traka & Mithen, 2009; Verkerk *et al.*, 2009). Bioactive compounds with antioxidant capacity such as phenolic compounds, ascorbic acid, vitamin E, carotenoids, and other plant secondary metabolites such as glucosinolates are well-known and recognized in their preventive roles against certain types of cancer and cardiovascular diseases (Cisneros-Zevallos, 2003; Scheerens, 2001). Figure 1, summarises the biosynthetic pathways of these compounds for brassica vegetables.

Nonetheless, there is still the need to identify the specific secondary metabolites of the different brassica crops and relate them to the alleged health benefits.

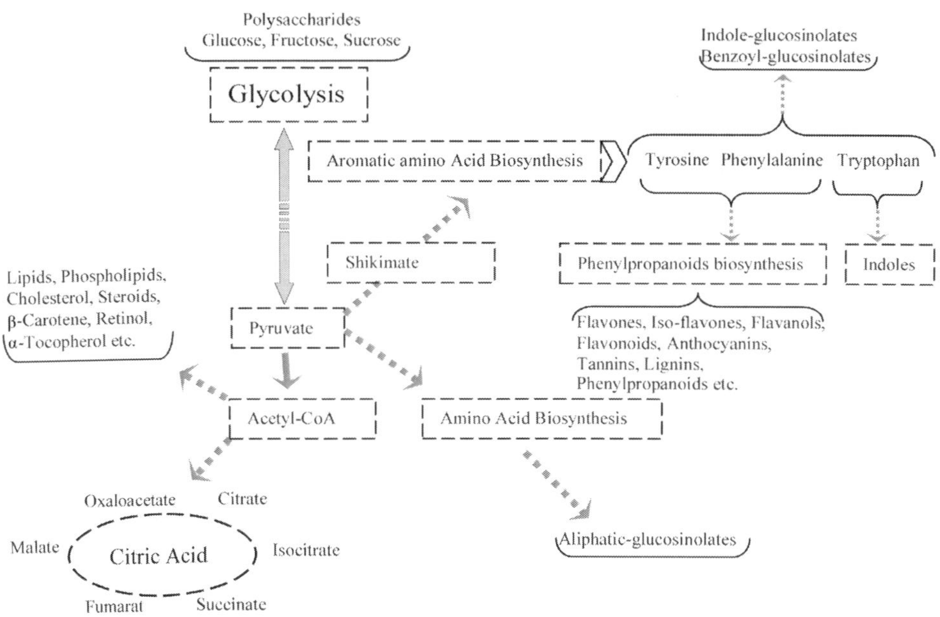

Figure 1. General biosynthesis pathways for Brassicaceae (Retrieved from Jahangir *et al.*, 2009a).

1.1. Glucosinolates

Glucosinolates (GLS) are secondary metabolites, characteristic for the *Caparales* order, and constitute the major class of these metabolites in brassica crops (Björkman *et al.*, 2011). The molecule comprises a β-thioglucoside N-hydroxysulphate, containing a side chain and a β-D-glucopyranose moiety (Sørensen, 1990). The structural diversity of GLS is mainly due to the variety of different substituents possible at the side-chain position R (Rosa *et al.*, 1997). Usually, GLS are divided into three chemical classes, depending on the respective amino acid precursor, aliphatic (methionine), indole (tryptophan) or aromatic (tyrosine or phenylalanine) (Giamoustaris & Mithen, 1996).

Glucosinolates do not reveal any bioactive role unless they are enzymatically hydrolysed to yield a variety bioactive breakdown products by myrosinase (thioglucoside glucohydrolase, E.C. 3.2.1.147), including isothiocyanates, nitriles, thiocyanates, oxazolidine-2-thiones and indolyl compounds (Grubb & Abel, 2006). The formation of these GLS breakdown products depend on several factors such as the specific GLS, Fe^{2+} availability, pH conditions, among others. For instance, the pH during hydrolysis determines the formation of either isothiocyanates (at physiological pH) or nitriles (at acidic pH) (Halkier & Du, 1997). The GLS breakdown products are of great concern due to their either beneficial or harmful properties. Among the beneficial uses are their anti-fungicidal and anti-bacterial properties, which create the natural protection of the plant itself with potential application to biofumigation (Fahey *et al.*, 2001, 2002; Angus *et al.*, 1994), and to humans as cancer-chemoprevention agents (Jahangir *et al.*, 2009a; Cartea & Velasco, 2008; Shapiro *et al.*, 2001; Rosa *et al.*, 1997).

Epidemiological evidences suggest that consumption of brassica vegetables reduces the risk for lung, stomach, colon and rectal cancers, most likely due to their glucosinolate content (Van Poppel *et al.*, 1999). As examples, it is known that isothiocyanates interfere with the mitochondria meditated apoptosis in cancer cells (Tang *et al.*, 2006), reduce the development of cardiovascular diseases, hypertension (Wu *et al.*, 2004), and also aiding in the skin protection against UV radiation (Talalay *et al.*, 2007). However, there are also epidemiological evidences that brassica consumption may lead to the incidence of other cancers, such as prostate (Giovannucci *et al.*, 2003). The biological mechanisms responsible for the harmful activity of GLS-derived compounds are only partly elucidated. From animal studies, it is known that isothiocyanates interfere with the synthesis of thyroid hormones, while thiocyanates compete with iodine and inhibit iodine uptake by the thyroid gland. In addition to the thyroid gland, main target organs are the liver, kidney and pancreas, showing altered weight and malfunction. The mechanisms for these phenomena are greatly unknown, although carcinogenic processes have been reported (Verkerk *et al.*, 2009).

GLS in brassica products, either in profile and concentration, can vary widely, depending on the cultivar, fertilization and environment (Ciska *et al.*, 2000). For instance, in a given plant species about 15 different GLS can be found from which four are present in significant amounts, and comparing *B. oleracea* with *B. rapa*, the first surely has grater GLS amounts and diversity (Verkerk *et al.*, 2009). Even though the three classes of GLS can be found in brassica vegetables, methionine-derived GLS are the most abundant (Mithen *et al.*, 2003) while indole GLS are in minority (Zukalová & Vašák, 2002). Glucosinolate composition of several brassica species is shown in Table 1.

Common to *B. oleracea* crops are glucobrassicin (3-indolylmethyl; precursor of ascorbigen) and glucoiberin (3-methylsulfinilpropyl; precursor of the iberin isothiocyanate). Sinigrin (2-propenyl; precursor of 4 aglycones: allyl isothiocyanate, allyl cyanide, 1-cyano-2,3-epithiopropane and allyl thiocyanate) is also found in a large majority of *B. oleracea* crops, particularly in kale and cabbage (Ciska *et al.*, 2000). Broccoli shows prevalence of glucoraphanin (4-methylsulfinylbutyl; precursor of sulphoraphane), above 50% of the total content as shown by Kushad *et al.* (1999), while Brussels sprouts and cauliflower exhibit high levels of sinigrin, progoitrin (2-hydroxy-3-butenyl) and glucobrassicin (Kushad *et al.*, 1999; Carlson *et al.*, 1987; VanEtten *et al.*, 1976). Among the GLS found in broccoli, the most bioactive and studied group are isothiocyanates, particularly sulforaphane (breakdown product from glucoraphanin), allyl isothiocyanate (breakdown product from sinigrin) and indole-3-carbinol (breakdown product from glucobrassicin) since these compounds were identified as having anti-cancer activity (Jones, Faragher, & Winkler, 2006).

Brassica rapa species shows small variation within GLS composition where characteristic GLS include gluconapin (3-butenyl), glucobrassicanapin (4-pentenyl), progoitrin (2-hydroxy-3-butenyl), gluconapoleiferin (2-hydroxy-4-pentyl) and gluconasturtiin (2-pentylethyl) (Padilla *et al.*, 2007, Rosa, 1997; Sones, Heaney, & Fenwick, 1984). While studying 33 *B. napus* L. leafy, forage, rutabaga, and oilseed crops, Velasco *et al.* (2008) found that even though aliphatic GLS were predominant either in seeds and leaves, indole GLS were more abundant in leaves. In seeds, progoitrin was found as the main glucosinolate in all crop groups while in leaves the characteristic GLS depended on crop. For forage and root crops progoitrin was more abundant whereas glucobrassicanapin was characteristic of oilseed and leafy crops. Also in oilseed rape, Bohinc *et al.* (2013) showed prevalence of sinalbin (4-hydroxybenzyl), glucobrassicin (3-indolmethyl) and progoitrin (2(R)-2-Hydroxy-3-butenyl).

There are sufficient evidences to support that the richness of GLS and respective breakdown products found in brassica vegetables have the potential to reduce the risk of cancer development in humans and therefore the respective comsumption should be increased.

1.2. Phenolic Compounds

Phenolic compounds are a large group of phytochemicals (more than 8000) widely dispersed throughout the plant kingdom and characterized by having at least one aromatic ring with one or more hydroxyl groups attached. Phenolics are produced in plants as secondary metabolites via the shikimic acid pathway. Phenylalanine ammonia-lyase (PAL, EC 4.3.1.5) is the key enzyme catalysing the biosynthesis of phenolics from the aromatic amino acid phenylalanine (Koukol & Conn, 1961).

Phenolic compounds have been reported to have multiple biological effects for human health, including anti-inflammatory, enzyme inhibition, antimicrobial, antiallergic, vascular and cytotoxic antitumor activity, but the most important action of phenolics is related to their antioxidant capacity (Wei, Miao, & Wang, 2011; de Pascual *et al.*, 2010; Podsedek, 2007; Cushnie & Lamb, 2005; Chu *et al.*, 2000; Podsedek *et al.*, 2000; Plumb *et al.*, 1997). Furthermore, phenolic compounds hold other properties such as hydrogen peroxide production in the presence of certain metals, the ability to scavenge electrophiles and inhibit nitrosation reactions and chelate metals and, therefore, they act by blocking the initiation of several human diseases (Fresco *et al.*, 2010; Pereira *et al.*, 2009). They are categorized into classes depending on their structure and subcategorized within each class according to the number and position of hydroxyl group and the presence of other substituents. The most widespread and diverse group of the polyphenols are flavonoids which are built upon C_6–C_3–C_6 flavone skeleton. In addition, other phenolic compounds such as benzoic acid or cinnamic acid derivatives have been identified in fruits and vegetables (Aherne & O'Brien, 2002; Robards *et al.*, 1999).

The nutritional interest of brassica crops is partly related to their diverse phenolic composition and related antioxidant capacity. These crops are generally rich in polyphenols, but the phenolic composition can be quite different among species and even among crops from the same species, qualitative and quantitatively (Velasco *et al.*, 2011; Francisco *et al.*, 2011, 2009; Ferreres *et al.*, 2005; Vallejo, Tomás-Barberán, & Ferreres, 2004; Llorach *et al.*, 2003; Nielsen *et al.*, 1993). The most widespread and diverse group of polyphenols in brassica species are flavonoids (mainly flavonols but also anthocyanins) and hydroxycinnamic acids, where flavonols such as quercetin and kaempferol, and anthocyanidins, show a greater efficacy as antioxidants, on a mole for mole basis, than the antioxidant nutrients ascorbic acid, vitamin E and carotenoids (Rice-Evans *et al.*, 1996; Rice-Evans *et al.*, 1995; Vinson *et al.*, 1995). Table 2 shows in summary the results of a recent study (Sikora *et al.*, 2012) concerning the phenolic composition of selected *B. oleracea* vegetables.

Table 2. Phenolic acids and flavonoids in some *Brassica oleracea* vegetables (mg/100 g fresh matter; Quantified from lyophilised samples after enzymatic hydrolysis by HPLC-DAD.). (Adapted from Sikora *et al.*, 2012)

	Caffeic acid	Ferulic acid	p-Coumaric acid	Sinapic acid	Quercetin	Kaempferol	Isorhamnetin	Apigenin	Luteolin
Broccoli	0.4	1.98	1.95	2.95	2.09	3.60	1.49	-	-
Brussels sprouts	3.98	2.28	4.24	13.57	0.59	1.40	-	-	-
Kale	25.95	5.76	14.49	13.39	8.48	26.90	-	-	-
Cauliflower	0.53	0.30	0.72	0.70	0.33	1.00	-	-	-
Green cauliflower	0.20	0.25	0.59	1.36	0.06	0.07	-	0.14	0.37

Among brassica crops, broccoli has been the most exhaustively studied with regard to polyphenol composition. Numerous and recent studies have shown that this crop (leaves, florets and sprouts) contains a high antioxidant potential linked to a high level of phenolic compounds (Moreno *et al.* 2006; Llorach *et al.*, 2003; Vallejo, Garcia-Viguera, & Tomás Barberán, 2003). Heimler *et al.*, (2006) compared the main phenolic compounds in several *B. oleracea* crops and stated that broccoli and kale varieties exhibit the higher phenolic content, particularly in regard to flavonoids. Quercetin, kaempferol and phenolic acids derivatives from the external and internal leaves, seeds and sprouts leaves of tronchuda cabbage have also been reported by several authors (Sousa *et al.*, 2007; Ferreres *et al.*, 2005; Sousa *et al.*, 2005) and the different composition seems to be conclusive for the antioxidant activity displayed by each plant part.

Anthocyanins have also been identified in brassica vegetables (Moreno *et al.*, 2010; Jahangir *et al.*, 2009b; Wu & Prior, 2005). For example, the red pigmentation of red cabbage, purple cauliflower and purple broccoli is caused by anthocyanins. The major anthocyanins identified in these crops are cyanidin derivatives. Cauliflower and red cabbage showed differences in their anthocyanin profiles: cyanidin-3, 5-diglucoside was absent in cauliflower, while it was well represented in red cabbage, together with the characteristic anthocyanin of the genus *Brassica*, cyanidin-3-sophoroside-5-glucoside. The p-coumaryl and feruloyl esterified forms of cyanidin-3-sophoroside-5-glucoside were predominant in cauliflower, while the sinapyl ester was mostly present in red cabbage (Jahangir *et al.*, 2009b; Lo Scalzo *et al.*, 2008). Red cabbage contains more than 15 different anthocyanins, which are acylglycosides of cyanidin (Dyrby, Westergaard, & Stapelfeldt, 2001). Seventeen different anthocyanins were present in broccoli (Moreno *et al.*, 2010) with the main peaks corresponding to cyanidin-3-O-digluco-side-5-O-glucoside acylated and double acylated with p-coumaric, sinapic, caffeic or ferulic acids. Several studies have shown that lignans (diphenolic compounds) are prevalent in the Brassicaceae family, and particularly in kale, broccoli and Brussels sprouts (Heinonen *et al.*, 2001), with lariciresinol and pinoresinol being the most abundant (Milder *et al.*, 2005). Even though a large variety of lignans are found, only a few percentage is converted into enterolignans, absorbed into the human body, which hold several biological activities, such as antioxidant and (anti) oestrogenic properties.

Further details on phenolic profiles in different *Brassica* species can be found in an extensive review by Podsedek (2007) and Cartea *et al.* (2011).

1.3. Ascorbic Acid

L-ascorbic acid (AA) is an odourless, white solid having the chemical formula $C_6H_8O_6$ and is a water-soluble vitamin. It is easily oxidized to form dehydroascorbic acid (DHAA), and thus oxidation is readily reversible from DHAA (Groff *et al.*, 1995). However, DHAA is unstable and it is spontaneously and enzymatically converted to 2,3-diketogulonic acid (Davey *et al.*, 2000) at physiological pH. The main biological functions of ascorbic acid can be defined as an enzyme cofactor, as a radical scavenger and as a donor/acceptor in electron transport either at the plasma membrane or in the chloroplasts. Ascorbic acid is able to scavenge the superoxide and hydroxyl radicals, as well as regenerate α-tocopherol and β-carotene through the reduction of the formed radicals (Davey *et al.*, 2000).

In addition to AA and DHAA, brassica vegetables include ascorbigen, which is formed as the result of the reaction between ascorbic acid and indolyl-3-carbinole, one of the degradation products of a glucosinolate, glucobrassicin. It is likely that some of the biological effects attributed to ascorbigen are mediated by its breakdown to ascorbic acid.

The water-soluble properties of ascorbic acid allow for the quenching of free radicals before they reach the cellular membrane. Ascorbic acid is important in collagen formation, thereby resulting in stabilization of the peptide. Indirectly, AA plays important regulatory roles throughout the entire body due to its involvement in the synthesis of hormones, hormone-releasing factors, and neurotransmitters (Jacoba, 1999; Groff *et al.*, 1995). L-ascorbic acid acts as a co-factor for at least eight hydroxylases and monooxygenases involved in synthesis of collagen, noradrenalin, serotonin and carnitine, as well as in detoxication of xenobiotics (Davies *et al.*, 1991). The participation in various free-radical processes is a likely cause of the known immuno-modulating and antiviral properties of AA (Uchide & Toyoda, 2011; Furuya *et al.*, 2008; Jariwalla *et al.*, 1996). It is widely used as vitaminous, regenerative and antiviral medication in the treatment of various respiratory viral infections including influenza, herpetic infections, viral hepatitis and other infectious diseases (Davies *et al.*, 1991; Parkinson, 1984).

The content of AA in brassica vegetables varies significantly among and within species: ascorbic acid levels varied over 4-fold in broccoli and cauliflower, 2.5-fold in Brussels sprouts and white cabbage and 2-fold in kale. The cause of these reported variations might be related to differences in genotype (Vallejo, Tomás-Barberán, & Garcia-Viguera, 2002; Kurilich *et al.*, 1999) and climatic conditions (Howard *et al.*, 1999).

Generally, white cabbage, one of the most popular brassica vegetables, is the poorest source of vitamin C among this group of crops (Podsedek, 2007), while broccoli, Brussels sprouts, cauliflower and cabbage generally show highest L-AA contents, ranging from 50 to >100 mg/100 g (Davey *et al.*, 2000), electing these crops as good sources of AA. Singh *et al.* (2007) found that AA content in cabbage (18 cultivars) ranged from 5.7 to 23.5 mg/100 g; cauliflower (2 cultivars) 13.8 and 24.8 mg/100 g; Brussels sprouts (2 cultivars), 14.6 and 17.0 mg/100 g. Chinese cabbage (4 cultivars) ranged from 5.62 to 12.6 mg/100 g, Broccoli: (6 cultivars) in broccoli ranged from 25.5 to 82.3 mg/100 g.

According to Gokmen *et al.* (2000), DHAA was the dominant form of vitamin C in cabbage, with 4-fold higher level than AA. In contrast to this report, Vanderslice *et al.* (1990) observed that the contribution of DHAA to the total vitamin C content was 14% or 8% in cauliflower and broccoli, respectively. These authors did not find DHAA in fresh cabbage. Those values were in agreement with that reported for broccoli by Vallejo, Tomás-Barberán, & Garcia-Viguera (2003a), who found that DHAA contribution to total vitamin C content was 11.3%.

1.4. Carotenoids

Carotenoids are lipophilic molecules, including a large group of >600 compounds, which are characterized by a polyisoprenoid structure, a long conjugated chain of double bond and a near bilateral symmetry around the central double bond, as common chemical features (Britton, 1995). The central chain (40-carbon basal structure) may carry cyclic end-groups which can be substituted with oxygen-containing functional groups. Based on their composition, carotenoids are divided in two classes, carotenes containing only carbon and hydrogen atoms, and oxocarotenoids (xanthophylls) which carry at least one oxygen atom. The pattern of conjugated double bonds in the polyene backbone of carotenoids determines their light absorbing properties and influences the antioxidant activity of carotenoids (Stahl & Sies, 2003). According to the number of double bonds, several cis/trans (E/Z) configurations are possible for a given molecule. Carotenoids tend to isomerize and form a mixture of mono- and poly-cis-isomers in addition to the all-trans form, which is predominant in nature.

The biological function of carotenoids has been widely researched and has shown a range of health protection effects. Much of this has been attributed to their antioxidant activity. β-carotene is the major precursor of vitamin A and other retinoids and also has effective reducing power and exhibit free radical scavenging properties. Some authors have shown anticarcinogenic effects (Bertram & Bortkiewicz, 1995; Toma *et al.*, 1995; Shklar & Schwartz, 1993) and these observations resulted in the test of this compound as a preventive medicine for certain types of cancers. Surprisingly, some intervention trials concerning the effects of β-carotene in humans found a higher relative risk for lung cancer in smokers who were given β-carotene (Omenn *et al.*, 1996; Blumberg, 1994).

Important dietary carotenoids found in brassica vegetables include β-carotene (both *cis* and *trans* isomers), lutein, zeaxanthin, cryptoxanthin, neoxanthin and violaxanthin (Podsedek, 2007; Muller, 1997; Wills & Rangga 1996; Hart & Scott, 1995; Heinonen *et al.*, 1989). In *B. oleracea*, kale is one of the vegetables with the highest carotenoid content (over 10 mg/100 g edible portion) (Muller, 1997), being Brussels sprouts intermediate (6.1 mg/100 g) and broccoli (1.6 mg/100 g), red cabbage (0.43 mg/100 g) and white cabbage (0.26 mg/100 g) low in total carotenoid content (Muller, 1997). Podsedek (2007) found the highest lutein + zeaxanthin content for kale (3.04–39.55 mg/100 g), being the amount of these compounds moderately high (0.78–3.50 mg/100 g) in broccoli and Brussels sprouts. Also in broccoli and Brussels sprouts as well as in green cabbage, in addition to lutein and trans-β-carotene, the presence of cis-β-carotene has been reported (Muller, 1997; Hart & Scott, 1995).

In another study, Singh *et al.* (2007) found that β-carotene content in cabbage (16 cultivars) ranged from 0.01 to 0.12 mg/100 g, where maximum contents were found in Quisto (0.12 mg/100 g) followed by Green Challenger and Rare Ball (0.11 mg/100 g). The lowest content of β-carotene was found in Pusa Mukta cultivar (0.01 mg/100 g). Lutein content was also recorded in the cabbage cultivars, ranging from 0.02 (Pusa Mukta) to 0.26 mg/100 g (Quisto).

In *B. rapa* species, 16 carotenoids were identified by Wills & Rangga (1996) in the chinensis, parachinensis and pekinensis subspecies, being lutein and β-carotene also the most abundant. In this study, Brussels sprouts ranked 11th (6.1 mg/100 g), broccoli 16th (1.6 mg/100 g), red cabbage 20th (0.43 mg/100 g), and white cabbage 21st (0.26 mg/100 g) based on total carotenoid content.

1.5. Vitamin E

Vitamin E is a term that includes a group of potent, lipid-soluble, chain-breaking antioxidants. Structural analyses have revealed that molecules having vitamin E antioxidant activity include tocopherols (α, β, γ, δ) and tocotrienols (α, β, γ, δ) (Schneider, 2005). The α-tocopherol form is the most abundant in nature highly accumulated in chloroplasts (DellaPenna & Pogson, 2006). It prevents lipid peroxidation by removal of singlet oxygen and lipid peroxyl radicals (Krieger-Liszkay *et al.*, 2008). The resultant tocopheroxyl radicals are reduced back to α-tocopherol by ascorbate trough the ascorbate–glutathione cycle.

In addition to vitamin E antioxidant activity and ability to scavenge free radicals, it is proposed that it may reduce the risk of cancer and prevent progression of precancerous lesions (Pinheiro-Sant'Ana *et al.*, 2011). Vitamin E also shows protective effects against the coronary heart disease due to inhibition of LDL oxidation (Stampfer & Rimm, 1995).

Although vegetables, in addition to fats, oils and cereal grains, constitute the major source of vitamin E in our diet, there are only few data of tocopherol content in vegetables. In general, the best sources of lipid-soluble antioxidants are kale and broccoli. Brussels sprouts have moderate levels of the above-mentioned compounds, while cauliflower and cabbage are characterized by their relatively low amounts. The descending order of total tocopherols and tocotrienols in brassica vegetables is as follows: broccoli (0.82 mg/100 g), Brussels sprouts (0.40 mg/100 g), cauliflower (0.35 mg/100 g), Chinese cabbage (0.24 mg/100 g), red cabbage (0.05 mg/100 g), and white cabbage (0.04 mg/100 g) (Piironen *et al.*, 1986). Kurilich *et al.* (1999) have also reported similar rank on the basis of concentration, but in their study total tocopherol values were about 2-fold higher. These differences are probably caused by the differing varieties and growing conditions. According to these authors, kale was the best source of α-tocopherol and γ-tocopherol (2.15 mg/ 100 g). Piironen *et al.* (1986) reported that α-tocopherol was predominant tocopherol in all brassica vegetables, except in cauliflower, containing predominantly γ-tocopherol. In contrast, Kurilich *et al.* (1999) reported lower concentration of γ-tocopherol than α-tocopherol in cauliflower. In *B. rapa* L. subsp. Sylvestris, Annunziata *et al.* (2012) reported that α-tocopherol and γ-tocopherol contents were of 4.0 ± 0.1 and 0.4 ± 0.1 μg/g (fw), respectively.

In various cabbage cultivars, Singh *et al.* (2007) found that Vitamin E (DL-α-tocopherol) content ranged from 0.03 (cv. Rare Ball) to 0.20 mg/100 g (cv. Green Cornell). The found DL-α-tocopherol values in that study are in agreement with the earlier report of Ching & Mohamed (2001) in which the DL-α-tocopherol value in cabbage, on fresh weight basis, was reported to be 0.69 mg/100 g.

2. Pre-, Postharvest and Processing Effects on the Bioactive Composition of Brassicas

Conscious of the health benefits from the consumption of brassicas, the focus is an examination of the current knowledge related to the effect of the complete chain (pre and postharvest procedures, storage, preservation and processing methods) on the content of the health-promoting compounds and its biological activity. An overview on the major mechanisms involved in the different processes (e.g. diffusivity, thermal degradation and others) responsable for the alteration can be a powerfull tool to assist in a optimization of the levels of these phytochemicals in the human diet.

2.1. Effects of Pre-harvest Conditions on Compounds of Brassica Vegetables

Brassicaceae biochemical composition, either in profile and concentration, varies widely depending on different factors such as cultivar, tissue types, environment situation fertilization, and other preharvest conditions.

Cultivars

Of the many hundreds of cruciferous species investigated, all are able to synthesize glucosinolates (Kjær, 1976). In this vegetable family, the plant's genetic background is one of the major factors determining GLS concentration and composition. At least 120 different GLS have been identifed in these plants, although closely related taxonomic groups typically contain only a small number of such compounds. Glucosinolates were evaluated in 5 groups and 65 accessions of *B. oleracea* (50 broccoli, 4 Brussels sprouts, 6 cabbage, 3 cauliflower, and 2 kale) grown under uniform cultural conditions by Kushad *et al.* (1999), concluding that GLS and their concentrations varied among the different groups and within each group. Among other authors, and as was present in the above section, Verkerk *et al.* (2009) summarised the GLS data present in the most economically important members. Considering the potential beneficial effects to human health, broccoli attracted attention after the discovery that it contains high levels of certain GLS with anticarcinogenic properties. However, among broccoli cultivars there is a large variation within the levels of these compounds (Rosa & Rodrigues, 2001; Hill *et al.*, 1987).

Regarding other phytochemical compounds, differences were also observed among cultivars. In a recent study, Samec *et al.* (2011) point to significant variability in total phenol and total flavonoid contents and antioxidant capacity between white and Chinese cabbage and also between two cultivars of Croatian white cabbage (Ogulinski and Varăzdinski). Singh *et al.* (2007) evaluated different cabbage cultivars, cauliflower, broccoli and brussels sprouts. These authors refered that, in comparasion, broccoli generally had the highest levels of phenolics (63.4 mg/100 gfw), ascorbic acid (52.9 mg/100 gfw), β-carotene (0.81 mg/100 gfw), lutein (0.68 mg/100 gfw) and α-tocopherol (0.47 mg/100 gfw), with brussels sprouts in a close second. Podsedek *et al.* (2006) compared several varieties of red cabbage, white cabbage (*B. oleracea* convar. *capitata*. var. *capitata*), savoy cabbage (*B. oleracea* convar. *capitata* var. *sabauda*), and brussels sprouts and conclued that highest content of phenolics was found in red cabbage (171 mg/100 gfw) and the lowest in white cabbage (21 mg/100 gfw). These kind of indications are important tools to identify plants with optimal health promoting potential and could be useful for producers who are trying to produce final products with added value. On the other hand, varieties with high content of health-promoting compounds enhance the possibility of studying the genes involved in its regulation and the feasibility of modifying profiles in specific plant genotypes.

Tissue Types

Phytochemical compounds vary greatly with tissue types of the plants (e.g. seeds, roots, leaves, shoots or flowers) and the developmental stages of the tissue. Azevedo & Rodriguez-Amaya (2005) stated that β-carotene and lutein of the kale samples had significantly higher levels in the mature leaves compared with the young leaves. The study done by Aleksander & Malgorzata (2010) has proved that during successive stages of maturation of rapeseed (*B. napus* L.), the content of phenolic acids varied in anatomical parts of the plant.

Brown *et al.* (2003) evaluated the glucosinolate content of various organs of the model plant *Arabidopsis thaliana* (L.) Heynh, and significant diferences were found among organs in both glucosinolate concentration and composition concluding that during seed germination and leaf senescence, there were significant declines in glucosinolate concentration. Other similar studies were made by Singh *et al.* (2007) in different cabbages, cauliflower, broccoli and Brussels sprouts, by Aleksander & Malgorzata (2010) in rapeseed (*B. napus* L), by Aires *et al.* (2011) in Portuguese kale, by Samec *et al.* (2011) in white and Chinese cabbage and by Cartea *et al.* (2012) in *B. rapa* crops. An analysis of 74 studies made by van Dam, Tytgat, & Kirkegaard (2009) allowed to conclude that roots have higher concentrations and greater diversity in GLS than shoots.

All of the raised relevant information related with plant tissue and its development suggest different biological functions of particular compounds that contributed to plants growth, play different roles in plant interaction with herbivores or pathogens and in the responses to environmental stresses (Gigolashvili *et al.*, 2009; van Dam *et al.*, 2009; van Dam & Raaijimakers, 2006). Therefore, the knowlegment of the role of these compounds in natural would provide insights to plant breeders wishing to manipulate composition of crop species in order to tackle the challenges of climate change and food insecurity.

Enviromental Factors

As mentioned, an important source of variation in the overall concentration of the health-related compounds are seasonal effects namelly, growing season, light and temperature and water supply (Ishii & Saijo 1987; Sang *et al.* 1986; Rosa, 1992).

Factors such as light intensity (irradiation), photoperiod, light quality (wavelength) and temperature affect plants´ physiological responses and thus may affect GLS content (Charron & Sam, 2004). Accordingly, winter or autumn seasons seem to induce lower GLS levels, due to short days, cool temperatures accompanied by frosts, and less radiation. However, synthesis and degradation of GLS compounds can occur in a wide range of climate conditions (Shattuck *et al.*, 1990). It seems that temperature stress induce higher GLS levels in brassicas, as was concluded by Rosa & Rodrigues (1998) in cabbage seedlings and Pereira *et al.* (2002) in two cultivars of broccoli sprouts. Nilsson *et al.* (2006) studied GLS levels in several cabbage varieties grown over a two-year period. Changes in total GLS levels were atributed to the increased of glucobrassicin which varied drastically between the two years of the study. The authors attributed the higher GLS levels in the second year to an increase of 1°C in average temperature. Schonhof *et al.* (2007a) evaluated GLS in broccoli grown in three different daily mean temperature (in the range 7.2 to 19.7 C) and two different daily radiation levels (in the range from 19 to 13.4 $mol.m^{-2}.day^{-1}$). The authors justify the different responses among GLS groups due to the various enzymes involved in GLS synthesis that were affected directly by temperature and radiation. Cartea *et al.* (2008) also reported variation in total glucosinolate content in a collection of 153 kales, 26 cabbages, and three Tronchuda cabbages varieties at different growing seasons grown in Spain and Portugal.

Several studies suggest that light quality can influence GLS content. Increased light intensity leads to higher production of these compounds as was shown for rape (Whitecross & Armstrong, 1972) and Brussels sprouts (Baker, 1974). However, the same authors suggest that, compared with temperature, ligh intensity may be of minor importance (Bjorkman *et al.*, 2011).

Environmental stress causes significant changes in crops bioactive composition. The stresses are numerous and often crop- or location-specific. They include a range of environmental factors such as radiation, water, high salinity, temperature, mineral nutrient deficiency, metal toxicity, air pollutants, and topography (Cogo *et al.*, 2011; Fortier *et al.*, 2010; McKenzie *et al.*, 2007; López-Berenguer, García-Viguera, & Carvajal, 2006; Rodovich *et al.*, 2005; Rangkadilok *et al.*, 2004; Vallejo, Tomás-Barberán, & Garcia-Viguera, 2003a; Champoliver & Merrien, 1996; Bennet & Wallsgrove, 1994; Zhao *et al.*, 1994). Although different stress factors may have different molecular targets, a common response to unfavorable environmental conditions is the occurrence of oxidative stress with increased levels of reactive oxygen species (ROS) (Grene, 2002; Smirnoff, 1995). Thus, stress modifies the secondary metabolite composition of plants (Jahangir *et al.*, 2009b), altering plant stress tolerance (Mittler, 2002) and the nutritional value of crop plants for the human diet (Verkerk *et al.* 2009; Jansen *et al.* 2008).

However, some environmental factors have a pronounced effect on brassicas phytochemical composition such as light. Light is known to regulate not only plant growth and development, but also the biosynthesis of both primary and secondary metabolites. Phenolic biosynthesis requires light or is enhanced by light, and flavonoid formation is absolutely light dependent and its biosynthetic rate is related to light intensity and density. However, different plants had a different response to light intensity alteration and the resulting total flavonoids and total phenolics production. The light effect was evaluated by Oh & Rajashekar (2009) in sprouts of alfalfa, broccoli and radish, by Pérez-Balibrea *et al.* (2008) in broccoli sprouts, and by Lefsrud, Kopsell, & Sams (2007) in kale.

UV-B (280–315 nm) is the most energetic radiation reaching the earth's surface. When plants are not acclimatised or are irradiated with UV-B levels above the current ambient radiation, this radiation can have detrimental effects on lipids, proteins and nucleic acids, and specifically affect the photosystem II by damaging its membranes and decreasing enzyme activities (Bassman, 2004). UV-B leads also to an inhibition of cell expansion by reducing levels of indole-3-acetic acid, thereby affecting plant morphology (Stratmann, 2003; Hollósy, 2002; Jansen *et al.*, 1998; Rozema *et al.*, 1997). The production of flavonoids and related phenolic compounds as a response to UV-B in several plant species is well documented (Sakalauskaitė *et al.*, 2012; Caldwell *et al.*, 2007; Fagerberg & Bornman, 2005; Jansen *et al.*, 2001; Mackerness, 2000; Caldwell *et al.*, 1999; Rozema *et al.*, 1997). However, similar studies in brassica vegetables are very limited being expetion Schonhof *et al.* (2007b) in broccoli, Gitz, Liu, & McClure (1998) in red cabbage, Kuhlmann & Muller (2009) and Sangtarash *et al.* (2009) in canola and Mewis *et al.* (2012) in broccoli.

Less is known about effects of UV-A (315–400 nm), however UV-A can induce the production of phenolics as was proved by Krizek *et al.* (1997) on cucumber and Krizek *et al.* (1998) on lettuce.

The response of plants to heat stress includes morphological alterations and anatomical modifications, as well as physiological and biochemical changes (Iglesias-Acosta *et al.*, 2010). Temperature influence the nature of epicular crystalline wax structures formed on leaf surface as was shown by Whitecross & Armstrong (1972) in rape, Baker (1974) and Reed & Tukey (1982) in leaves of brussels sprout. High temperature (>15 ºC) lowers the levels of lutein and β-carotene in broccoli (Schreiner *et al.*, 2012; Schonhof *et al.*, 2007b). Richards *et al.* (2008) observed a positive correlation between the daily maximum temperature and the α-tocopherol concentration in Canola (*B. napus*).

Reduced water content alters the chemical composition of plants, which can influence their tolerance to insect herbivory (Ahuja, Rohloff, & Bones, 2010; Khan, Ulrichs, & Mewis, 2010) and could lead to changes on phytochemical content (Cogo *et al.*, 2011; Khan, Ulrichs, & Mewis, 2011). For instance, in the case of broccoli, rapeseed and turnip root, less water supply caused increases in glucosinolate content as observed by Paschold *et al.* (2000), Bouchereau *et al.* (1996) and Zhang *et al.* (2008), respectively. As a consequence of the reduced water content in crops, nutrient concentration is increased, and is particularly relevant in crops that are usually consumed at a high water concentration.

All these plant interactions with environmental stress factors including, light, temperature and water content, are known to lead to the activation of various defense mechanisms resulting in a qualitative and/or quantitative change in plant metabolite production. Pre-harvest conditions are known to affect plants produce signaling molecules that cause a direct or indirect activation of metabolic pathways. These conditions affect the production of phytochemicals, such as carbohydrates (sucrose and glucose), amino acids, phenolics and glucosinolates. The effect of these stress on the metabolism of Brassicaceae are all well explained in Jahangir *et al.* (2009b).

Concerns about the effect of environmental factors, in particular light, temperature and irradiation, have arisen in the last decades because the stratospheric ozone layer has been depleted, leading to increased levels of solar radiation reaching the Earth's surface. The threat to ensuring productivity in global agriculture and horticulture due to ozone depletion and loss of plant species cannot be overstated nor should it be overlooked (Menis *et al.*, 2012). Therefore, future work is required on detailed effects of these factors on the biocomposition of brassica vegetables, driven by the hope of improving crop yield in afflicted areas.

Fertilization

Fundamental differences between organic and conventional production systems, particularly in soil fertility management, influences crops nutritional composition. The fertilization and mineral nutrient application level, type and value, directly influence the level of nutrients available in plants and indirectly influence plant physiology and the biosynthesis of secondary metabolites or phytonutrients (Picchi *et al.*, 2012; Brandt *et al.*, 2011; Lo Scalzo *et al.*, 2008; Martínez-Ballesta *et al.*, 2008).

Conventional farming utilises fertilisers that contain soluble inorganic nitrogen and other nutrients, which are easily available to plants. Variations in the phytochemical levels of cauliflower can be caused by nitrogen fertilizer (Lisiewska & Kmiecik, 1996). The total glucosinolate level was also observed to increase as a response to sulphur availability in turnip rape (*B. rapa*) (Kim *et al.*, 2002) and kale (*B. oleracea* L. Acephala Group) (Kopsell *et al.*, 2003), while three broccoli cultivars showed an increase in total glucosinolate content at the start of the inflorescence development followed by a rapid decrease depending on its fertilization with sulphur (Vallejo *et al.*, 2003).

Extreme agronomic conditions (rich sulphur fertilization) enhanced the phenolic content of three cultivars from freshly harvested broccoli inflorescences var. *italica*. The cultivars growning under rich fertilisation showed also higher ascorbic acid content than those grown under the poor fertilisation (Vallejo *et al.*, 2003).

Influence of nitrogen and sulphur fertilization on quality of canola (*B.napus* L.) were also evaluated by Ahmad *et al.* (2007). The authors stated that glucosinolate content increased from 13.6 to 24.6 μmol/g as the S concentration was increased from 0 to 30 kg/ha and the highest tested N concentration resulted in the highest glucosinolate contents (19.9 μmol/g). Another fertilizer, selenium, when applied up to a certain doses induced higher glucosinolates levels, particularlly in sulforaphane, while higher doses decreased glucosinolate production (Hsu *et al.*, 2011; Robbins *et al.*, 2005). Submitting broccoli to salt stress increased their glucosinolate content, indicating the involvement of these compounds in a stress response (Lopez-Berenguer *et al.*, 2008). Some exceptions are also reported, as in the case of cadmium stress which produced no change in glucosinolate production in *B. rapa* (Siemens *et al.*, 2002).

Organic vegetable production is characterized for relying more on natural mechanisms of growth, yield and disease control. Application of compost to sandy soil increased organic matter, cation exchange capacity, available nutrients and biological properties (Lombardi-Boccia *et al.*, 2004). Organic fertilization has a stimulatory effect on phenolic accumulation in broccoli florets. The higher concentrations of phenolics found in florets can be explained by the role of organic fertilizers in inducing the acetate shikimate pathway, resulting in higher production of phenolics such as flavonoids as was evaluated by Sousa *et al.* (2008) in tronchuda cabbage. The accumulation of indole, aliphatic and aromatic glucosinolates could be enhanced by the presence of low nitrogen and high sulphur fertilizers. For example the use of a sulphur fertilizer produced an increase in the glucosinolates, gluconapin, sinigrin and progoitrin on mustard (*B. juncea* L.) seeds (Kaur *et al.*, 1990).

The nutrition value response of brassica vegetables to different agricultural pratice could be accomplished directly by changing fertilisation regimes. Such manipulations may allow plant breeders and biotechnologists to significantly modify the sensorial, anti-cancer or biofumigant properties of numerous brassica crops.

2.2. Effects of Postharvest Conditions on Compounds of Brassica Vegetables

There are many ways in which produce is treated after harvest. *Brassica* vegetables generally undergo a variety of postharvest operations which include cooling, washing, sorting and grading, packaging and storage at ambient or chill temperatures, until sale and use by consumer. Storage conditions after harvesting are essential to maintain product quality and changes in GLS levels and other bioactive compounds of brassica under various storage conditions have not been systematically studied until recently. The effect of storage conditions on brassica vegetables quality embrace normally its duration, temperature, relative humidity and atmosphere conditions, being broccoli the main vegetable evaluated.

Time and Temperature

Lower temperature decreases metabolic rates and thereby slows down deterioration. The duration of storage is of course also of major importance, since the concentration of bioactive compounds change over time. However, the effect of these postharvest factors on GLS contents on broccoli appear to be contradictory. Some authors reported an increase on GLS content (Verkerk, Dekker, & Jongen, 2001; Hansen *et al.*, 1995), others a decrease (Force *et al.*, 2007; Vallejo, Tomás-Barberán, & Garcia-Viguera, 2003b; Rodrigues & Rosa, 1999) and still others reported no effects (Winkler *et al.*, 2007; Rangkadilok *et al.* 2002). The observed discrepancies could be explain by the fact that the authors evaluate different GLS compounds, use different storage conditions, different methodology of evaluation and different broccoli varieties and origins. For example, Rodrigues & Rosa (1999) found a decrease in glucoraphanin content of broccoli cv 'Tokyodome' stored at 4 °C for 5 days, whilst Winkler *et al.* (2007) reported no significant loss on broccoli cv "Marathon" stored at 4 °C during 28 days. Jones, Faragher, & Winkler (2006) in their review stated that the most important postharvest conditions necessary for maintaining broccoli quality is low temperature (<4 C). This condition maintains cellular integrity and in the process appear to maintain GLS content by preventing the mixing of glucosinolates with myrosinase.

In other brassica vegetables, Shattuck, Kakuda, & Shelp (1991), stated that low temperature altered the concentration of several GLS in peeled root and peel tissues, but did not induce a qualitative change in the GLS profile in rutabaga (*B. napus* spp. *rapifera* (Metzg.) Sinsk). Also, Hwang, Jang, & Kim (2012) suggest that the total glucosinolate content in Korean cabbage did not significantly changed over 7 days of storage at 4 °C, but reported small decreases in this content at ambient temperature. Similar results were found by Chong & Berard (1983) in cabbage while Ki-Hwan *et al.* (1992) reported a gradual decreases in GLS content. The decrease of glucosinolates may occur by hydrolysis that can take place caused by senescence (Dekker, Verkerk, & Jongen, 2000).

Cold storage had no effect on the antioxidant capacity, total phenols or flavonol content, but reduced the content of ascorbic acid and soluble sugars on curly kale (*B. oleracea* L. var. *acephala*) (Hagen *et al.*, 2009). Generally vegetables show a rapid loss of ascorbic acid at ambient temperature (Favell, 1998). In fact, several studies indicate that high storage temperate combined with longer storage periods contribute to losses in AA in different brassica vegetables (Pramanik *et al.*, 2004; Lee & Kader, 2000; Gil *et al.*, 1999; Gil *et al.*, 1998; Barth & Zhuang, 1996). The effect of ambient and refrigerated storage temperature on post-harvest phytochemical compounds (ascorbic acid, chlorophyll, β-carotene and total antioxidant activity) of broccoli florets was investigated during storage by Nath *et al.* (2011) and observed significant losses during storage for up to 144 h under refrigerated conditions.

Leja *et al.* (2001) observed stability of the total polyphenol content in broccoli during a 7 day storage at 5 ºC. Vallejo, Garcia-Viguera, & Tomás-Barberán (2003) demonstrated major losses of total flavonoids (62–59%), but not of AA, in broccoli inflorescences stored for 7 days at 1 °C (cold storage simulation), followed by a 3-day period at 15 °C, mimeting a retail period. Conversely, Winkler *et al.* (2007) showed that extended low temperature storage (1 or 4 ºC for up to 28 days), did not influence flavonoid content in broccoli inflorescences.

As for carotenoids, contents of lutein in broccoli increase when kept at daily mean temperature between 7 and 13 °C (Schonhof *et al.*, 2007). The extractable β-carotene, lutein as well as tocopherol contents in Korean cabbage were decreased by increasing storage period either at 4 °C or ambient temperature (Hwang, Jang, & Kim, 2012). In short, longer storage periods (> 7 days) at ambient temperature reduce the levels of health promoting compounds.

Relative Humidity

Relative humidity can influence water loss, decay development, the incidence of some physiological disorders, and uniformity of vegetable ripening. In spite of the influence of RH on the post-harvest quality losses in vegetables, few studies have been published on the direct effect of RH in GLS and other bioactive compounds. In general, high RH of 98–100% is recommended to maintain postharvest quality in brassica vegetables (Toivonen & Forney, 2004; Wang, 2003). RH only appears to be a critical factor in GLS retention when postharvest temperature rise above approximately 4 °C (Rangkadilok *et al.*, 2002; Rodrigues & Rosa, 1999). The decrease in glucoraphanin coincided with a marked loss of visual quality (i.e. yellowing), indicating probable loss of membrane integrity and mixing of glucosinolates with myrosinase. However, when broccoli was stored at 4 °C, there was no difference in glucoraphanin content after 7 days in either open boxes at ambient humidity (approximately 60% RH) or in plastic bags (approximately 100% RH) (Rangkadilok *et al.*, 2002).

Atmosphere Packaging

The amount of research dedicated to controlled atmosphere (CA), modified atmosphere (MA) and modified atmosphere packaging (MAP) on brassica vegetables has historically lagged behind that of fruits. Most of the research carried out with these vegetables has been focused on a few crops such as broccoli.

Package gas composition determines the effectiveness of any packaging system. Reduced O_2 and elevated CO_2 concentrations must be sufficiently stringent to slow metabolism and provide shelf-life extension while also being within the tolerance range of the stored commodity to avoid induction of anaerobic conditions. Temperature management is critical since gas composition within packages change with temperature. Most of the research on CA, MA and MAP involve empirical observations of changes in various quality factors over time in experiments that involve placing a product in several combinations of gas atmospheres and also different temperatures. Duration in the package after sealing is also important, as commodity tolerance to atmospheres may change over time. Reports of MAP experiments in which a vegetable product is placed in a number of packages are not particularly useful unless the packaging materials are selected since they are expected to create a particular desired gas composition. Finally, various processes of produce ripening and senescence do not have the same O_2 and CO_2 optima for maximizing beneficial responses.

In the last years, there has been some reports concerning suitable atmospheres for broccoli storage and to preserve its nutritive value, in particular the levels of health promoting compounds. Rangkadilok *et al.* (2002) found that the level of GLS, in particular glucoraphanin, was significantly higher with a CA treatment (1.5% O_2 and 6% CO_2) compared to regular atmosphere during 25 days storage. CA treatments with elevated CO_2 (21% O_2 + 10% CO_2, and 21% O_2 + 20% CO_2) and air treatments (21% O_2) were found to increase the glucoraphanin content over the first 5 days of storage at 5 °C. On the other hand, CA treatments with reduced O_2 concentration (1% O_2, 1% O_2 + 10% CO_2) led to a steady decrease of glucoraphanin content in broccoli (var. *italica*) during 20 days at 5 °C (Chao-Jion *et al.*, 2006). The study of Schouten *et al.* (2009) describes the effects of controlled atmosphere (1.5 kPa O_2 and 15 kPa CO_2) and temperature on GLS levels in broccoli (cv. '1997'). Storage under these conditions showed to maintain GLS content for at least 14 days. Fernández-León *et al.* (2013) found that broccoli (var. *italica*) stored under controlled atmospheres (10% O_2 and 5% CO_2) at 20 °C, maintained the contents of phenolic compounds and GLS for 2 and 4 days, respectively. Modified and controlled atmosphere approaches (1–2% O_2 and 5–10% CO_2) in combination with temperatures of 1-5 °C and a high relative humidity (98-100% for all) have shown to preserve phytochemical compounds of broccoli, namely carotenoids and ascorbic acid (Jacobsson *et al.*, 2004; Barth & Zhuang, 1996).

Rangkadilok *et al.* (2002) used MAP with low-density polyethylene (LDPE) bags and atmospheres of approximately 3% O_2 and 11% CO_2 at 4 °C to maintain glucoraphanin levels of broccoli var. *italica* during 10 days. In comparison with the GLS content of freshly harvested broccoli, glucoraphanin content of 'Marathon' broccoli heads stored for 7 days at 1 °C under MAP using also LDPE bags, decreased by approximately 48% (Vallejo *et al.*, 2003). Jones, Faragher, & Winkler (2006) concluded that CA and MAP appear to be useful tools for maintaining GLS levels but further work is needed to understand the involved mechanisms. Broccoli (var. *italica*) packaged with three different MAP had prolonged storability up to 28 days with high quality attributes and health-promoting compounds, ascorbic acid and total phenolic compounds, compared with unwrapped control broccoli (Serrano *et al.*, 2006).

Schreiner *et al.* (2007) found that MAP packaging (1% O_2 and 21% CO_2) is suitable to maintain GLS levels of mixed packaged broccoli and cauliflower for 7 days at 8 °C. All three MAP treatments evaluated by Cheng-Guo *et al.* (2009) slowed the decrease rates of individual, total aliphatic and indole glucosinolates contents in broccoli florets var. *italica* when compared to those in the control.

Only a few studies are available regarding the influence of atmosphere packaging storage conditions on polyphenolic and GLS compounds in other brassica vegetables. The influence of different initial phenolic contents in pak choi (*B. campestris* L. ssp. *chinensis* var. *communis*) submited to controlled storage atmosphere conditions of 1.5–2.5% O_2 and 5–6% CO_2 or normal air atmospheres at 2 °C and 99% relative humidity were evaluated by Harbaum-Piayda *et al.* (2010). The level of flavonoids increased more in controlled atmospheres than in normal air, but hydroxycinnamic acids were unaffected. Mamphol *et al.* (2010) evaluated the effect of modified atmosphere packaging on the quality and bioactive compounds of Chinese cabbage (*B. rapa* L. ssp. *chinensis*). The created atmosphere (2% O_2

and 7% CO_2) inside of one type of biorientated polypropylene packaging improved the overall appearance, moderately maintained chlorophyll a and b, and the bioactive compounds and antioxidant scavenging activity, and remained marketable for up to 10 days at 10 °C. The authors conclued that gas composition within the packages influenced the retention of bioactive compounds as well as the overall quality.

Glucosinolate content of cauliflower heads (cv. Freemont) in air and/or controlled-atmosphere storage (3% O_2 and 5% CO_2) storage at 0 °C were evaluated by Hodges *et al.* (2005). No differences in glucosinolate profiles were found between storage treatments. However, the authors stated that the glucosinolates gluconapin and glucobrassicin increased for each treatment during storage, albeit later in controlled-atmosphere conditions. Glucobrassicin was the major glucosinolate component, and the dramatic increase in concentration was reflected in the total glucosinolate levels of air-stored cauliflower on day 28 of storage. Levels of the other glucosinolates did not change during storage but glucoiberin content decreased after day 28. The found increases in the levels of gluconapin and glucobrassicin could be related to metabolic changes associated with natural and/or stress-induced senescence.

Therefore, from the described studies, it can be concluded that both CA storage and MAP appear to be a useful tools to maintain the contents of bioactive compounds after harvest in brassica vegetables. However, further investigation is needed to clearly elucidate the atmospheres that may best maintain the different compounds.

2.3. Effects of Industrial and Home Processing on Compounds of Brassica Vegetables

Minimal Processing

The term minimally processed vegetables is applied to any fresh vegetable that has been physically altered from its original form, but remains in a fresh state (Gomez-Lopez *et al.*, 2008). Regardless of the commodity, it has been trimmed, peeled, washed, and cut into a 100% usable product that is subsequently bagged or pre-packaged (IFPA, 2009). Normally, modified atmosphere packaging (MAP), is used to increase the shelf-life of these vegetables (Murcia *et al.*, 2003).

The preparation operations involved in the postharvest chain of these products will trigger complex reactions mechanisms, physical and physiological processes which lead to changes in the levels of bioactive compounds. These processing steps could be expected to reduce a rapid enzimatic depletion of several natural antioxidants. The cutting operations may also increases the exposure of antioxidant compounds to oxygen.

Normally, brassica vegetables are usually chopped up before consumption. Cutting the fresh plant tissues creates optimal conditions for myrosinase, so a high degree of glucosinolate hydrolysis can be expected, and in the extreme case, pulping of plant tissues results in the complete breakdown of glucosinolates by autolysis (Mithen *et al.*, 2000; Rodrigues & Rosa, 1999; Rosa *et al.*, 1997; Daxenbichler, 1991). For instance, the initial concentration of GLS (62 μmol/100 g fw) in broccoli florets dropped by 75% after 6 hours at 25 ºC and only 50% of these were hydrolyxed to isothiacyonates (Song & Thornalley, 2007). However, as indicated, GLS levels do not necessarily decline rapidly after this operation and even induction can take place. Verkerk *et al.* (1997) observed elevated levels of all indolyl and some aliphatic GLS after chopping and prolonged exposure to air of different *Brassica* vegetables, something which could have large influence on quality factors such as flavour and anticarcinogenicity (Koritsas *et al.*, 1991). Other authors reported that removing part of the stalk during preparation of fresh-cut broccoli product caused a statistically significant increase (30 to 40%) in the concentration of GLS (Martinez *et al.*, 2007). This increse in extracted GLS could be due to GLS synthesis induction as a response to cutting (Verkerk, Dekker, & Jongen, 2001) or again to an increae on the percentage of inflorescences in the cut product, where the GLS are present in significant greater concentrations (Rangkadilok *et al.* 2002). However, and in resume, two opposed mechanisms can occur, which may justify the unpredictable effects on the damage tissue: a) the reduction in glucosinolates attributed to plant myrosinase-mediated hydrolysis of glucosinolates at the cut surfaces; b) the stress-induced synthesis of glucosinolates (Mithen *et al.*, 2000).

With regard to ascorbic acid and in the case of broccoli, cutting reduced vitamin content of the fresh-cut product by 27% with respect to the uncut product. However, the decrease was justified by the different content of ascorbic acid present in the plant parts that were analiyzed (De Ancos *et al.*, 2011). One of the physiological responses to mechanical damage is the increase in PAL activity which leads to the accumulation of synthetised phenolic compounds protecting the plant tissues from water loss and attacks by pathogenic microrganisms (Reyes *et al.*, 2007).

In relation to other bioactive compounds, such as carotenoids and tocopherol compounds, it seems that the different steps followed to produce fresh-cut broccoli did not alter its content or bioavailability in the product (Granado-Lorencio *et al.*, 2008).

Sanitation treatments are often performed by immersion of the cut products in washing oxidizing solutions (chemical sanitizers). Consequently, in fresh-cut vegetables the reduction of oxidative compounds is expected to occur both by leaching and oxidation phenomena. Several studies with fresh-cut vegetables and fruit demonstrated variable losses of ascorbic acid, ranging between 20 and 60% (mangoes; Gonzalez-Aguilar *et al.*, 2008; peppers, Howard & Hernandez-Brenes, 1998; red sweet peppers, Raffo *et al.* 2008). Although few studies have been undertaken in sanitation treatments effects on GLS content, a study done by Martinez *et al.* (2007) showed that sanitation treatments with or without sodium hypochlorite did not significantly influence GLS levels in cut broccoli florets stored for 23 days at 4 °C. However, as for water-soluble glucosinolates, the respective levels can also be affected by washing conditions during the preparation of brassica vegetables. For example, washing brassica after cutting with hot water or with water for longer time periods promotes the loss of glucosinolates (Benner *et al.*, 2003). In accordance with Martinez *et al.* (2007), the different sanitation treatments used (water and a 150 ppm sodium hypoclorite solution) did not affect broccoli ascorbic acid content.

Other sanitizing systems still under study and used to obain a safe fresh-cut vegetable can influence the bioactive value of these products. For example, the combined effect of ionizing radiation and MAP can improve antioxidant activity in some products such as chinese cabbage (*B. rapa*) if appropriate treatment conditions are selected (Tomás-Callejas, 2012; Ahn *et al.* 2005). Also in fresh-cut broccoli, the combination of 24 $\mu molm^{-2}s^{-1}$ intensity light exposure with 7 °C storage temperature maintained quality (antioxidant power and total phenols) and extended its shelf-life (Zhan *et al.*, 2012). The effects of UV radiation processing on several bioactive compounds (anthocyanins, total phenolics, lycopene, ascorbic acid, chlorophylls, antioxidants enzymes, etc.) of plant produce have been reviewed by Alothman *et al.* (2009).

The effects of neutral electrolysed water (NEW), ultraviolet light C (UV-C) and superatmospheric O_2 packaging, single or combined, on the quality of fresh-cut kailan-hybrid broccoli for 19 days at 5 ºC were studied by Martínez-Hernández *et al.* (2013). Combining treatments achieved increases in the activities of antioxidant enzymes while mantaining fatty acid composition. Nonetheless, further investigations are required to better optimise conditions, and preserve produce overall quality.

Heat Treatments

Domestic and industrial processing usually involves thermal treatment. Moreover the use of heat is usually required prior to brassica vegetable consumption. During heating many mechanisms take place: thermal degradation of GLS and breakdown products, enzymatic breakdown of GLS, myrosinase inactivation and leaching of GLS and breakdown products into cooking water (Dekker, Verkerk, & Jongen, 2000; Rodrigues & Rosa, 1999). There is a great amount of literature available concerning the effects of different thermal treatment normally used, namely blanching, sterilization, cooking in water, vapor or microwaves on the content, composition, antioxidant activity and bioavailability of antioxidants in different brassica vegetables, particular broccoli.

In general, blanching is a heat treatment needed to stabilize frozen or dry vegetables through the inactivation of given enzymes that can affect products quality during storage or prior to further processing such as heat sterilization (Gonçalves *et al.*, 2009). However, heat-sensitive nutrients may be lost and, in water blanching, soluble constituents may be leached, resulting in large volumes of effluent. The quality of blanched products depends significantly on the time and temperature of blanching and also on the physical and chemical properties of vegetable to be blanched.

Many authors reported significant losses in the content of ascorbic acid and polyphenols and decreases in antioxidant activity in brassica vegetables after blanching (cauliflower and broccoli, Lisiewska & Kmiecik, 1996; cauliflower and cabbage, Puupponen-Pimia *et al.*, 2003; broccoli, Zhang & Hamauzu, 2004 and Gębczyński & Lisiewska, 2006; turnip greens, Mondragón-Portocarrero *et al.* 2006; Brussels sprouts, Czarniecka-Skubina, 2002 and Olivera *et al.*, 2008; kale, Korus, 2011; turnip greens, Martinez *et al.*, 2013). Amin & Wee Yee (2005) showed that when blanching brassica vegetables for 15 min, the decrease in antioxidant activity compared with the raw material was 40% in Chinese cabbage but only 4% in red cabbage.

Volden *et al.* (2008) observed significant losses in blanched red cabbage on the levels of glucosinolates, polyphenols and anthocyanins. Also, blanching, reduced total aliphatic and indole GLS by 31% and 37%, respectively. L-ascorbic acid (L-AA), total phenols (TP), anthocyanins were on average reduced by 19, 15, 38%, respectively, in different varieties of cauliflower (Volden, Bengtsson, & Wicklund (2009). Wennberg *et al.* (2006) investigated the effects of blanching in two cultivars of shredded white cabbage. After 5 min of blanching the total GLS levels had been decreased substantially in two tested cultivars by 50 and 74%. The individual GLSs were affected to different degrees. Cieslik *et al.* (2007) investigated the effects of blanching in several different vegetable, finding a reduction by 2–30% for total GLS levels. Oliviero *et al.* (2012) evaluated the effect of water content and temperature on glucosinolate thermal degradation in broccoli (*Brassica oleracea* var. *italica*). The authors found that degradation could be described by first-order kinetics for all glucosinolates and all tested water contents. In the temperature range 60-100 °C the sample with a 13% water content showed the lowest degradation rate, whereas at 120 °C the degradation rate increased with the water content.

Oerlemans *et al.* (2006), focused the attention on thermal degradation of GLS in red cabbage. The authors blanched the samples by microwaving (without adding water) in order to inactivate the myrosinase and subsequently heated the samples at different temperature/time combinations in the absence of additional water to exclude leaching losses. The chemical degradation was described by first order kinetics and Arrhenius-like temperature dependency. The parameter estimation showed that at temperatures below 110 °C indole glucosinolates have a significantly higher degradation rate constant compared to aliphatic glucosinolates.

It is believed that steam blanching would protect broccoli from the loss of phytochemicals in comparison to seeping broccoli in boiling water. However, steam blanching of broccoli decreased ascorbic acid concentration about 30% (Howard *et al.*, 1999). Murcia *et al.* (2000) observed that losses of AA content in broccoli were 50-51% in florets and 54-55% in steams independently of blanching time (1 to 2.5 min).

In most studies on the effect of thermal processing, the total loss of GLS is the result of many mechanisms occurring simultaneously. Leaching of GLS into the cooking water has been predicted by simulations to be the major factor responsible for GLS losses during boiling of vegetables (Verkerk *et al.*, 2009; Verkerk, 2002; Verkerk, Dekker, & Jongen, 2001;). Chevolleau *et al.* (1997) and Chevolleau *et al.* (2002) reported a 10% degradation of glucobrassicin after heating for 1 h at 100 ºC and observed the formation of a new breakdown product, 2-(30 -indolylmethyl). Rosa & Heaney (1993) found that boiling Portuguese cabbage for 10 min was sufficient to reduce the total glucosinolate content by more than 50%. Ciska & Kozlowska (2001) cooked white cabbage for 30 min and observed the highest decrease after 5 min of cooking (35%), which gradually decreased to a 87% loss after 30 min.

The effect of other thermal processing on GLS content and antioxidant compounds has been investigated in many brassica vegetables. Cooking methods were shown to affect the contents of nutrient and health-promoting compounds such as ascorbic acid, carotenoids, polyphenols, and glucosinolates in Brussels sprouts, white and green cauliflower, broccoli, curly kale and different varieties of cabbage (Gawlik-Dziki, 2008; Sikora *et al.*, 2008; Cieslik *et al.*, 2007; Rungapamestry *et al.*, 2007; Rungapamestry *et al.*, 2006; Lin & Chang, 2005; De Sa & Rodriguez-Amaya, 2004; Wu *et al.*, 2004; Zhang & Hamauzu, 2004; Ciska & Kozlowska, 2001; Rosa & Heaney, 1993).

The effects of five domestic cooking methods, including steaming, microwaving, boiling, stir-frying, and stir-frying followed by boiling (stir-frying/boiling), on the nutrients and health-promoting compounds of broccoli were investigated by Yuan *et al.* (2009). The results show that total aliphatic and indole glucosinolates were significantly modified by all cooking treatments but not by steaming. In general, steaming led to the lowest loss of total glucosinolates, while stir-frying and stir-frying/boiling, the most popular methods for most homemade dishes in China, showed the highest loss revealing steaming the best cooking method to maintain broccoli nutrients.

The results obtained by Gliszczyńska-Swigło (2006) also indicated that steam-cooking of broccoli results in an increase in polyphenols, as well as in the main glucosinolates and their total content as compared with fresh broccoli, whereas cooking in water had the opposite effect. Steam-cooking of broccoli had no influence on ascorbic acid, whereas cooking in water significantly lowered its content. Both, water- and steam-cooking of broccoli results in an increase in β-carotene, lutein, and α- and γ-tocopherols as compared with fresh broccoli.

Microwave cooking is thought to be an eficient alternative for cooking vegetables due to the low amount of cooking water required, therefore, it is expected that limited leaching of nutrients occurs (Czarniecka-Skubina, 2002). However, microwaving cooking of broccoli with water is shown to decrease in antioxidant components and total carotenoids in broccoli (Lopez-Berenguer *et al.*, 2007; Zhang & Hamauzu, 2004). Also, Vallejo, Tomás-Barberán, & García-Viguera (2002), stated that microwaving broccoli resulted in 40% loss of ascorbic acid and a 74% loss of glucosinolates. In contrast to these results, Verkerk & Dekker (2004) observed a 78% increase in total extractable glucosinolate content of red cabbage after microwave cooking for 4.8 min a 900 W, as well as the inactivation of myrosinase.

Martínez-Hernández *et al.* (2013) evaluated the nutritional quality changes of kailan-hybrid broccoli before and after industrial boiling, steaming, sous vide (SV), microwaving (MW), SV-MW and grilling throughout 45 days at 4 °C. Apparently, cooking increased the total phenolic content up to 2.0 and 1.7-fold for grilling and MW, respectively, owing to a better extraction. SV–MW, SV and MW produced the highest total antioxidant capacity increase (around 5.4–4.7-fold), contrary to the low enhancements of boiling and grilling (2.9-fold). The total carotenoid content was enhanced by boiling. Also, Song & Thornalley (2007) stated that cooking by steaming, microwaving and stir-frying did not produce significant losses in GLS whereas boiling showed significant losses due to compound leaching into the cooking water in broccoli, Brussels sprouts, cauliflower and green cabbage. Most of the glucosinolates losses (approximately 90%) were detected in the cooking water. Similarly, different authors reported enhancement of total carotenoids content, total phenolic content and antioxidant capacity after cooking, microwaving or steaming of brassica vegetables (Roy *et al.*, 2009; Miglio *et al.*, 2008; Podsędek, 2007; Halvorsen *et al.*, 2006; Turkmen, Sari & Velioglu, 2005; Zhang & Hamauzu, 2004). These increases were probably due an enhanced molecule extractability since a partial destructuration of the vegetable tissues is observed when cooking.

Canning is one of the main methods used by the food industry to preserve seasonally available vegetables. However, is also the most severe heat treatment. Jaworska, Kmiecik, & Maciejaszek (2001) pointed out that the nutritive value of canned vegetables is significantly reduced, and Hunter & Fletcher (2002) found that the antioxidant activity in sterilized vegetables is lower than in frozen products. Czarniecka-Skubina (2002) stated that in regard to ascorbic acid content, canning of Brussels sprouts was the worst preservation method, decreasing this content about 66%. Still, in the case of canned broccoli only 16% of original ascorbic acid was retained (Murcia *et al.*, 2000). Oerlemans *et al.* (2006) observed that canning red cabbage results in significant GLS thermal degradation (73%) confirming the earlier observations of Dekker & Verkerk (2003) in cabbage.

Freezing

Normally, consumers use frozen vegetables, mainly for convenience, time-saving and practical reasons (Ninfali & Bacchiocca, 2003). The low temperatures commonly used for frozen foods can maintain initial quality and nutritive value practically unchanged, making freezing one of the most efficient and adequate preservation methods to extend vegetable quality.

Alteration in bioactive compounds during freezing operation and frozen storage is mainly caused by the damage to vegetable tissue by growth of water crystals and disintegration of cells – especially when the freezing process takes a long time (Gonçalves *et al.* 2011; Sosińska & Obiedziński, 2011). However, there is limited research regarding the influence of the freezing process on glucosinolate content and other health related compounds in brassica vegetables. Song & Thornally (2007) found loss of about 33% on total GLS in various brassicas (broccoli, Brussels sprouts, cauliflower and green cabbage), while Quinsac *et al.* (1994) found almost a complete degradation of GLS in sprouts of sea kale. Total GLS presented a high loss rate during cold storage of broccoli, mainly due to decrease of the major GLS present in broccoli inflorescences namely glucoraphanin, glucobrassicin, and neoglucobrassicin (Valllejo *et al.*, 2003). Moreover, Rodrigues & Rosa (1999) determined exactly the same amount of glucosinolates in broccoli flowers (after blanching, freezing and 5 days storage at −20 °C) as in raw vegetables (where the procedure of determination involved lyophilisation and grinding).

The process of freezing kale leaves did not significantly reduce the level of analysed antioxidants or their antioxidant activity. After one year of storage, the total content of polyphenols in frozen kale was on average 12% lower than that found in frozen products directly after freezing (Korus & Lisiewska, 2011). Some authors showed that the effect of long-term freezer storage on the total phenolic content was minimal in comparison to blanching (Puupponen-Pimiä *et al.*, 2003). Similar results have been obtained for Brussels sprouts by Czarniecka-Skubina (2002) and Gębczyński & Lisiewska (2006) and Sikora *et al.* (2008) in frozen broccoli. The freezing operation also did not change the level of AA, which is also stable in frozen broccoli and cauliflower during a 12-month storage at -25 °C (Gonçalves *et al.*, 2011). Under similar conditions, AA content decreased by 3–18% for broccoli and 6–13% for cauliflower (Favell, 1998; Lisiewska & Kmiecik, 1996). Higher losses were observed for cabbage, which lost 30% of AA after storage under similar conditions (Puupponen-Pimia *et al.*, 2003).

From a wider perspective, the requirement to better understand the role and fate of a range of natural and process induced phytochemicals on both brassica vegetables stability and human health suggests that considerable areas of research remain to be explored.

3. Consumer Oriented Strategies to Increase Brassica Vegetable Intake

The health benefits of a rich diet in brassica vegetables in particular, and in fruits and vegetables (F&V) in general, have been recognized for some time due to its abundant content on phytochemicals. Inappropriate nutrition of these products is a significant causative factor for many chronic diseases currently afflicting developed countries, namely cancers, hypertension and heart disease. As a result, national and international health organizations have focused increasing effort in recent years on defining and promoting healthy diets. For example, the World Health Organization (2003; 2002) reports that the consumption of up to 400 g per day of F&V could reduce the total worldwide burden of disease by 1.8%, and reduces the burden of ischaemic heart disease and is chaemic stroke by 31% and 19%, respectively. In practice, the consumption of F&V of at least 5 portions a day is one of the most important strategies of to public health in relation to nutrition. In relation cruciferous vegetables, the recommendation points out of including at least 2-3 times per week as part of your diet, and make the serving size at least 1-1/2 cups.

Despite all these benefits, people do not properly follow the minimum recommended (Blanck *et al.,* 2008). Generally understand is that healthy eating promotes well-being and reduces the risk of incidence of certain diseases (Povey *et al.,* 1998). According to a recent survey from the Food Standards Agency, 50% of participants now know the recommendations for F&V intakes and yet, for example, currently the average UK consumption of F&V is 3 portions per day (FSA, 2002). So, if awareness is not the key factor in motivating people to eat healthily, there must be some other cause. Indeed, Dibsdall *et al.* (2002) point out some other factors – such psychosocial or lifestyle factors as higher barriers than the normally justification of access to F&V or affordability.

Many studies have revealed variables that influence F&V consumption among elderly populations worldwide (Riediger *et al.*, 2008; Yeh *et al.*, 2008). In general, the main factors that contribute to F&V intake are: demographic factors like age and gender (Rasmussen *et al.*, 2006; Ricciuto *et al.*, 2006; Reime *et al.*, 2000, Anderson *et al.*, 1992), psychological factors (Kristal *et al.*, 1995); socioeconomic class (Smith & Baghurst, 1992) and lifestyle behaviour. Studies have shown that people of higher socioeconomic classes have healthier and more nutritionally balanced diets than those of lower socioeconomic classes (Riediger *et al.* 2008, Ricciuto *et al.* 2006). Several studies find that, in terms of F&V consumption: men consume less than women (Dosil-Diaz *et al.*, 2008, Baker & Wardle, 2003; Perez, 2002; Thompson *et al.* 1999), smokers consume less than non-smokers (Holick *et al.*, 2002; Perez, 2002; Thompson *et al.*, 1999), and singles consume less than married people (Riediger *et al.*, 2008; Nepal *et al.*, 2011). For example, Baker & Wardle (2003) found that females consume more F&V than males, which they attribute to the poorer nutritional knowledge of males. The authors also found that males are less likely to know the recommended F&V intake and the benefits associated with F&V consumption.

However, others authors raise the problem in two different ways. First, many adults are unaware of what they are eating and think their diet is healthy when it is not (Glanz *et al.*, 1997; Lechne *et al.*, 1997). A number of factors may be acting to raise misunderstandings and to reduce confidence in nutrition communications. Messages about diet and health have often changed over time as a result of increases in scientific knowledge, increased complexity, and also variety of sources become increasingly complex and finally, many sources of health messages (government agencies, health-related organizations, food industry, and consumer groups). Second, most consumers purchase vegetables based on their sensory preferences, like colour and taste, many times as opposed to perceived nutrition or health value (Cox *et al.*, 2012; Glanz *et al.*, 1998; Drewnowski, 1996). Yet, it is far difficult to convince the consumers on appropriate vegetables preparation conditions if the sensory preferences are not met. However in certain vegetables, like brassicas, bitterness, an undesirable taste might be a positive feature, allowing consumers to select broccoli sprouts with the highest glucosinolate content (Fahey *et al.* 1997 and 1998). Also, in some cases, farmers and food industry measure glucosinolate content merely as a way of predicting excessive bitterness of Brussels sprouts and therefore meeting the consumer taste, whereas some scientists propose enhancing glucosinolates in broccoli sprouts for better health (Drewnowski & Gomez-Carneros, 2000; Fenwick *et al.*, 1990). When it comes to bitter phytonutrients, the demands of good taste and good health may be wholly incompatible.

Thus it is important to explore the consumer behaviour and the motives behind. Once consumer sensory preferences and motives behind the behaviour are known, negotiation between the consumer's choices and appropriate preparation conditions for phytochemicals can be achieved, where consumer is easily motivated to use the information on vegetable handling practices and new vegetable preparation conditions. Therefore health education strategies to encourage individuals to choose healthy food must be developed to promote a better diet. Moreover, the consumer is more aware of the link between diet and health, more concerned about self-care and personal health (Toner & Pitman, 2004), and is seemingly demanding more information on how to achieve better health through diet.

Thereafter, the question "How do we help consumers increase their F&V consumption?" continues without a single and simple answer. Tailoring or personalizing nutrition education messages has been shown to be effective in promoting dietary change (Perez, 2002; Ammerman *et al.*, 2001; Ciliska *et al.*, 2000; Campbell *et al.* 1994). In the United States, the 5 A Day for Better Health Program is one of the best-recognized health promotions aimed at increasing the consumption of F&V. It was promoted heavily through a public–private partnership. However, newer recommendations encompassed in the latest version of the US Department of Agriculture's food guide, MyPyramid, represent an increase from the older 5 A Day for Better Health recommendation and specify certain F&V groups for consumption (Stables *et al.*, 2005; Sorensen *et al.*, 1999). Some of these intervention trials have shown effects that are significant but not clinically important or sustainable, many at high cost (Devine *et al.,* 2005; Ciliska *et al.*, 2000). The high cost and limited availability of tests in community highlight a need to other interventions that can be implemented. Benner *et al.* (2003) propose the use of conceptual model to gather and disseminate information essential for successful products, given the example of eating broccolis. With the chain information model (CIM) the authors have developed a consumer-orientated tool for expert teams within production chains to improve products consumption in an efficient and effective way. Azagba & Sharaf (2011) suggest the need for a multifaceted approach for example, through the media and other community-organized nutrition programs. The varied nutrition messages should be picked up by the communication media, which are increasingly involved in disseminating scientific information to the public. Television news and print media has been distinguished by its alleged influence on disclosure of the effects of F&V on human health. A new and evolving area in the promotion of dietary behavioural change is 'e-learning', the use of interactive electronic media to facilitate teaching and learning on a range of issues including health. E-learning has grown out of recent developments in information and communication technology, such as the internet, interactive computer programs, interactive television and mobile telephones. The high level of accessibility, combined with emerging advances in computer processing power, data transmission and data storage, makes interactive e-learning a potentially powerful and cost-effective medium for improving dietary behaviour (Kerr *et al.*, 2012; Harris *et al.*, 2011).

Other original intervention is the Garden-based nutrition-education programs. This program for youth are gaining in popularity and are viewed by many as a promising strategy for increasing preferences and improving dietary intake of F&V and increased willingness to taste these products (Robinson-O´Brien, Story, & Heim, 2009). However, empirical evidence in this area is relatively scant. Therefore, there is a need for well-designed, evidenced-based, peer-reviewed studies to determine program effectiveness and impact.

Winning consumers is based also on delivering satisfaction and quality. Therefore convenience is important, whether it be in conveniently packaged healthy snacks or ease in meal preparation, thus challenging the food industry to design and produce an array of food products that are perfectly tuned to the wishes of individual consumers. New packaging materials and packaging techniques like modified atmosphere and controlled atmosphere, coatings, improved chilling techniques, and new mild preservation techniques provide possibilities for foods with longer shelf-life and, at the same time, improved sensorial characteristics (Rooij, 2000; Galizzi & Venturini, 1996). Also, current food labelling regulations allow claims to be made on packaging, which notify consumers of the amounts of healthy bioactive compounds found within the product and provides information to help consumers make educated choices about incorporating F&V into a healthy diet including tips, recipes, and interactive tools.

Conclusion

The Brassicaceae family includes a wide range of horticultural crops, some of them with economic significance and extensively used in the diet throughout the world. *Brassica* species are a rich source of health promoting compounds. In this chapter, the significance of health-related compounds such as glucosinolates, phenolics and other antioxidants, and also the influence of environmental conditions and processing procedures on brassica vegetables were reviewed.

However, more studies are needed to evaluate the effects of the multiple interacting factors that influence the different compounds levels under varying chain conditions. Such data sets will provide a wealth of information to growers, producers and consumers. Integration of the results from global scale methodology, where different plant compounds are studied at an increasingly higher sensitivity and their environmental interactions, is needed. This information will be of great value for identifying mechanisms for plant resistance, identifying and increasing the health promoting effects of brassica vegetables and predicting the productive and processing effects on the vegetables.

Therefore, for improving the health-quality, production and consumption of these vegetables, incentive programs are necessary in order to enhance the antioxidant potential of our daily food supply.

References

Aherne, S.A., O'Brien, N.M., (2002). Dietary flavonols: Chemistry, food content, and metabolism. *Nutrition* 18, 75–81.

Ahmad, G., Jan, A., Arif, M., Jan, M.T., Khattak, R.A., (2007). Influence of nitrogen and sulfur fertilization on quality of canola (*Brassica napus* L.) under rainfed conditions. *J Zhejiang Univ Sci B.* 8(10), 731-737.

Ahn, H.J., Kim, J.-H., Kim, J.-K., Kim, D.-H., Yook, H.-S., Byun, M.-W., (2005). Combined effects of irradiation and modified atmosphere packaging on minimally processed Chinese cabbage (*Brassica rapa* L.). *Food Chem.* 89(4), 589-597.

Ahuja, I., Rohloff, J., Bones, A.M., (2010). Defense mechanisms of Brassicaceae: implications for plant-insect interactions and potential for integrated pest management. A review. *Agron Sustain Dev.* 30, 311–348.

Aires, A., Fernandes, C., Carvalho, R., Bennett, R.N., Saavedra, M.J., Rosa, E.A., (2011). Seasonal effects on bioactive compounds and antioxidante capacity of six economically importante brássica vegetables. *Molecules* 16(8), 6816-6832.

Aleksander, S., Malgorzata, N.K., (2010). Studies on influence of rapeseed vegetation stages on level of phenolic compounds. *Journal of Oilseed Brassica* 1(1), 12-18.

Alothman, M., Bhat, R., Karim, A.A., (2009). UV radiation-induced changes of antioxidant capacity of fresh-cut tropical fruits. *Innov Food Sci Emerg Technol* 10, 512-516.

Amin, I., Wee Yee, L., (2005). Effect of different blanching times on antioxidant properties in selected cruciferous vegetables. *J. Sci. Food Agric.* 85, 2314–2320.

Ammerman, A., Lindquist, C., Hersey, J., Jackman, A.M., Gavin, N.I., Garces, C., Lohr, K.N., Cary, T.S., Whitener, B.L., (2001). Efficacy of Interventions to Modify Dietary Behavior Related to Cancer Risk. Summary, Evidence Report/Technology Assessment: Number 25. Rockville, Md: Agency for Healthcare Research and Quality; February 2001. AHRQ Publication 01-E029.

Anderson, A., Hunt, K., (1992). Who are the 'healthy eaters'? Eating patterns and health promotion in the west of Scotland. *Health Educ J* 51(1), 3-10.

Angus, J.F., Gardner, P.A., Kirkegaard, J.A., Desmarchelier, J.M., (1994). Biofumigation: isothiocyanates released from *Brassica* roots inhibit growth of the takeall fungus. *Plant Soil* 162, 107-112.

Annunziata, M.G., Attico, A., Woodrow, P., Oliva, M.A., Fuggi, A., Carillo, P., (2012). An improved fluorimetric HPLC method for quantifying tocopherols in *Brassica rapa* L. subsp. sylvestris after harvest. *J Food Comp Anal* 27(2), 145-150,

Azagba S., Sharaf, M.F., (2011). Disparities in the frequency of fruit and vegetable consumption by socio-demographic and lifestyle characteristics in Canada. *Nutrition Journal* 10, 118

Azevedo, C.H., Rodriguez-Amaya, D.B., (2005). Carotenoid composition of kale as influenced by maturity, season and minimal processing. *J. Sci. Food Agric.* 85(4), 591–597.

Baker, A.H., Wardle, J., (2003). Sex differences in fruit and vegetable intake in older adults. *Appetite* 40(3), 269-75.

Baker, E.A., (1974). The influence of environment on leaf wax development in *Brassica oleracea* vat. *gemmifera*. *New Phytol* 73, 955-966.

Barth, M.M., Zhuang, H., (1996). Packaging design affects antioxidant vitamin retention and quality of broccoli florets during postharvest storage. *Postharvest Biol Technol* 9(2), 141–150.

Bassman, J.H., (2004). Ecosystem consequences of enhanced solar ultraviolet radiation: secondary plant metabolites as mediators of multiple trophic interactions in terrestrial plant communities. *Photochem. Photobiol.* 79, 382–398.

Benner, M., Geerts, R.F.R, Linnemann, A.R., Jongen, W.M.F., Folstar, P., Cnossen, H.J., (2003). A chain information model for structured knowledge management: towards effective and efficient food product improvement. *Trends Food Sci. Technol.* 14(11), 469–477.

Bennett, R.N., Wallsgrove, R.M., (1994). Secondary metabolites in plant defence mechanisms. *New Phytol.* 127, 617–633.

Bertram, J.S., Bortkiewicz, H., (1995). Dietary carotenoids inhibit neoplastic transformation and modulate gene expression in mouse and human cells. *Am. J. Clin. Nutr.* 62, 1327S-1336S.

Björkman, M., Klingen, I., Birch, A.N.E., Bones, A.M, Bruce, T.J.A., Johansen, T.J., Meadow, R., Mølmann, J., Seljåsen, R., Smart, L.E., Stewart, D., (2011). Phytochemicals of Brassicaceae in plant protection and human health - Influences of climate, environment and agronomic practice. *Phytochemistry* 72(7), 538-556.

Blanck, H.M., Gillespie, C., Kimmons, J.E., Seymour, J.D., Serdula, M.K., (2008). Trends in fruit and vegetable consumption among US men and women, 1994- 2005. *Prev Chronic Dis* 5:A35.

Blumberg, J., Block, G., (1994). The Alpha-Tocopherol, Beta-Carotene Cancer Prevention Study in Finland. *Nutr Rev.* 52(7), 242-245.

Bohinc, T., Hrastar, R., Košir, I.J., Trdan, S., (2013). Association between glucosinolate concentration and injuries caused by cabbage stink bugs Eurydema spp. (Heteroptera: Pentatomidae) on different Brassicas. *Acta Sci. Agron.* [online]. 35(1), 1-8.

Bouchereau, A., Clossais-Besnard, N., Bensaoud, A., Leport, L., Renard, M., (1996). Water stress effects on rapeseed quality. *Eur. J. Agron.* 5, 19–30.

Brandt, K., Leifert, C., Sanderson, R., Seal, C.J., (2011). Agroecosystem Management and *Nutrition*al Quality of Plant Foods: The Case of Organic Fruits and Vegetables. *Crit Rev Plant Sci* 30, 177–197.

Britton, G. (1995). Structure and properties of carotenoids in relation to function. *FASEB J.* 9, 1551-1558.

Brown, P.D., Tokuhisa, J.G., Reichelt, M., Gershenzon, J., (2003). Variation of glucosinolate accumulation among different organs and developmental stages of *Arabidopsis thaliana. Phytochemistry* 62, 471–481

Caldwell, M.M., Bornman, J.F., Ballaré, C.L., Flint, S.D., Kulandaivelu, G., (2007). Terrestrial ecosystems, increased solar ultraviolet radiation, and interactions with other climate change factors. *Photochem Photobiol Sci* 6, 252–266.

Caldwell, M.M., Searles, P.S., Flint, S.D., Barnes, P.W., (1999). Terrestrial ecosystem responses to solar UV-B radiation mediated by vegetation, microbes, and abiotic photochemistry. *In*: Physiological Plant Ecology (eds. Press, M.C., Schole, J.D., Barker, M.G.), pp. 241-262. Blackwell, London.

Campbell, M.K., DeVellis, B.M., Strecher, V.J., Ammerman, A.S., DeVellis, R.F., Sandler, R.S., (1994). Improving Dietary Behavior: The Effectiveness of Tailored Messages in Primary Care Settings. *Am J Public Health* 84(5), 783–787.

Cao, G., Sofic, E., Prior, R.L., (1996). Antioxidant capacity of tea and common vegetables. *J. Agric. Food Chem.* 44, 3426–3431.

Carlson, D.J., Daxenbichler, M.E., VanEtten, C.H., Kwolek, W.F., Williams, P.H., (1987). Glucosinolates in crucifer vegetables: Broccoli, brussels sprouts, cauliflower, collards, kale, mustard greens, and kohlrabi. *J. Amer. Soc. Hort. Sci.* 112, 173–178.

Cartea, M.E., de Haro, A., Obregón, S., Soengas, P., Velasco, P., (2012). Glucosinolate variation in leaves of *Brassica rapa* crops. *Plant Foods For Human Nutrition (Dordrecht, Netherlands)* 67(3), 283-288.

Cartea, M.E., Francisco, M., Soengas, P., Velasco, P., (2011). Phenolic Compounds in *Brassica* Vegetables. *Molecules* 16, 251-280.

Cartea, M.E., Velasco, P., (2008). Glucosinolates in *Brassica* foods: bioavailability in food and significance for human health. *Phytochem Rev.* 7, 213-229.

Cartea, M.E., Velasco, P., Obregón, S., Padilla, G., de Haro, A., (2008). Seasonal variation in glucosinolate content in *Brassica oleracea* crops grown in northwestern Spain. *Phytochemistry* 69, 403–410.

Champolivier, L., Merrien, A., (1996). Effects of water stress applied at different growth stages to *Brassica napus* L. var. *oleifera* on yield, yield components and seed quality. *Eur. J. Agron.* 5(3-4), 153-160.

Chao-Jiong, X, De-Ping. G., Yuan, J., Gao-Feng, Y., Qiao-Mei, W., (2006). Changes in glucoraphanin content and quinone reductase activity in broccoli (*Brassica oleracea* var. *italica*) florets during cooling and controlled atmosphere storage. *Postharvest Biol Technol* 42(2), 176-184.

Charron, C., Sams, C.E., (2004). Glucosinolate Content and Myrosinase Activity in Rapid-Cycling *Brassica Oleracea* Grown in a Controlled Environment. *J. Amer. Soc. Hort. Sci.* 129, 321-330.

Cheng-Guo, J., Chao-Jiong, X., Wei, J., Yuan, J., Gao-Feng, Y., Bing-Liang, W. Qiao-Mei, W., (2009). Effect of modified atmosphere packaging on visual quality and glucosinolates of broccoli florets. *Food Chem.* 114, 28–37.

Chevolleau, S., Debrauwer, L., Boyer, G., & Tulliez, J., (2002). Isolation and structure elucidation of a new thermal breakdown product of glucobrassicin, the parent indole glucosinolate. *J Agric Food Chem* 50, 5185–5190.

Chevolleau, S., Gasc, N., Rollin, P., Tulliez, J., (1997). Enzymatic, chemical and thermal breakdown of H3 -labeled glucobrassicin, the parent indole glucosinolate. *J Agric Food Chem* 45, 4290–4296.

Ching, L.S., Mohamed, S., (2001). Alpha-tocopherol content in 62 edible tropical plants. *J Agric Food Chem* 49, 3101–3105.

Chong, C., Bérard, L.S., (1983). Changes in glucosinolates during refrigerated storage of cabbage. *J. Amer. Soc. Hort. Sci.* 108:688- 691.

Chu, Y.-F., Sun, J., Wu, X., Liu, R.H., (2002). Antioxidant and antiproliferative activities of common vegetables. *J. Agric. Food Chem.* 50, 6910–6916.

Chu, Y.H., Chang, C.L., Hsu, H.F., (2000). Flavonoid content of several vegetables and thcir antioxidant activity. *J. Sci. Food Agric.* 80, 561-566.

Cieślik, E., Leszczyńska, T., Filipiak-Florkiewicz, A., Sikora, E., Pisulewski, P.M., (2007). Effects of some technological processes on glucosinolate contents in cruciferous vegetables. *Food Chem.* 105(3), 976-981.

Ciliska, D., Miles, E., O'Brien, M.A., Turl, C., (2000). Effectiveness of communitybased interventions to increase fruit and vegetable consumption. *J Nutr Educ.* 32(6), 341-352.

Ciska, E., Kozlowska, H., (2001). The effect of cooking on the glucosinolates content in white cabbage. *Eur Food Res Technol* 212, 582–587.

Ciska, E., Martyniak-Przybyszewska, B., Kozlowska, H., (2000). Content of glucosinolates in cruciferous vegetables grown at the same site for two years under different climatic conditions. *J Agric Food Chem.* 48(7):2862-2867.

Cisneros-Zevallos, L., (2003). The Use of Controlled Postharvest Abiotic Stresses as a Tool for Enhancing the Nutraceutical Content and Adding-Value of Fresh Fruits and Vegetables. *J. Food Sci.* 68(5), 1560-1565.

Cogo, S.L.P., Chaves, F.C., Schirmer, M.A., Zambiazi, R.C., Nora, L., Silva, J.A., Rombaldi, C.V., (2011). Low soil water content during growth contributes to preservation of green colour and bioactive compounds of cold-stored broccoli (*Brassica oleraceae* L.) florets. *Postharvest Biol Technol.* 60(2), 158-163.

Cohen, J., Kristal, R., Stanford, J., (2000). Fruit and vegetable intakes and prostate cancer. *J. Natl. Cancer Inst.* 9, 61–68.

Cox, D.N., Melo, L., Zabaras, D., Delahunty, C.M., (2012). Acceptance of health-promoting *Brassica* vegetables: the influence of taste perception, information and attitudes. *Public Health Nutr.* 15(8), 1474-1482.

Cushnie, T.P.T., Lamb, A.J., (2005). Antimicrobial activity of flavonoids. I*nt. J. Antimicrob. Agents* 26, 343-356.

Czarniecka-Skubina, E., (2002). Effect of the material form, storage and cooking methods on the quality of Brussels sprouts. *Pol. J. Food Nutr. Sci.* 11/52(3), 75–82.

Davey, M.W., Van Montagu, M., Inze, D., Sanmartin, M., Kanellis, A., Smirnoff, N., Benzie, I.J.J., Strain, J.J., Favell, D., Fletcher, J., (2000). Plant L-ascorbic acid: chemistry, function, metabolism, bioavailability and effects of processing. *J. Sci. Food Agric.* 80, 825-860.

Davies, M.B., Partridge, D.A., Austin, J.A., (1991). Vitamin C: Its Chemistry and Biochemistry, Royal Society of Chemistry, Cambridge.

Daxenbichler, M.E., Spencer, G.F., Carlson, D.G., Rose, G.B., Brinker, A.M., Powell, R.G., (1991). Glucosinolate composition of seeds from 297 species of wild plants. *Phytochemistry* 30, 2623–2638.

De Ancos, B., Sánche-Moreno, C., Plaza, L., Cano, M.P., (2011). Nutritional and Health Aspects of Fresh-Cut Vegetables. *In*: Advances in Fresh-Cut Fruits and Vegetables Processing, Martin-Belloso, O., Soliva-Fortuny, R. (Eds.), 2011, CRC Press, pp 146-184.

de Pascual-Teresa, S., Moreno, D.A., Garcia-Viguera, C., (2010). Flavanols and Anthocyanins in Cardiovascular Health: A Review of Current Evidence. *Int. J. Mol. Sci.* 11, 1679-1703.

De Sa, M.C., Rodriguez-Amaya, D.B., (2004). Optimization of HPLC quantification of carotenoids in cooked green vegetables—comparison of analytical and calculated data. *J Food Comp Anal* 17, 37–51.

Dekker, M., Verkerk, R., (2003). Dealing with variability in food production chains: a tool to enhance the sensitivity of epidemiological studies on phytochemicals. *Eur J Nutr.* 42(1), 67-72.

Dekker, M., Verkerk, R., Jongen, W.M.F., (2000). Predictive modelling of health aspects in the food production chain: a case study on glucosinolates in cabbage. *Trends Food Sci. Technol.* 11, 174–181.

DellaPenna, D., Pogson, B.J., (2006). Vitamin Synthesis in Plants: Tocopherols and Carotenoids. *Ann. Rev. Plant. Biol.* 57, 711-738.

Devine, C.M., Farrell, T.J., Hartman, R., (2005). Sisters in Health: Experiential Program Emphasizing Social Interaction Increases Fruit and Vegetable Intake among Low-Income Adults. *J Nutr Educ Behav.* 37, 265-270.

Dibsdall, L.A., Lambert, N., Bobbin, R.F., Frewer, L.J., (2002). Low-income consumers' attitudes and behaviour towards access, availability and motivation to eat fruit and vegetables. *Public Health Nutr.* 6(2), 159–168

Dosil-Diaz, O., Ruano-Ravina, A., Gestal-Otero, J.J., Barros-Dios, J.M., (2008). Consumption of fruit and vegetables and risk of lung cancer: a case-control study in Galicia, Spain. *Nutrition* 24(5), 407–413.

Drewnowski, A., (1996). From asparagus to zucchini: mapping cognitive space for vegetable names. *J Am Coll Nutr* 15, 147–153.

Drewnowski, A., Gomez-Carneros, C., (2000). Bitter taste, phytonutrients, and the consumer: a review. *Am J Clin Nutr* 72(6), 1424-1435

Dyrby, M., Westergaard, N., Stapelfeldt, H., (2001). Light and heat sensitivity of red cabbage extract in soft drink medel systems. *Food Chem* 72, 431–437.

Fagerberg, W.R., Bornman, J.F., (2005). Modification of leaf cytology and anatomy in *Brassica napus* grown under above ambient levels of supplemental UV-B radiation. *Photochem Photobiol Sci.* 4(3), 275-279.

Fahey, J.W., Stephenson, K.K., Talalay, P., (1998). Glucosinolates, myrosinase, and isothiocyanates: three reasons for eating brassica vegetables. *In*: Shibamoto T, Terao J, Osawa T, eds. Functional foods for disease prevention I. Fruits, vegetables, and teas. Vol 1. Washington, DC: American Chemical Society, 1998:16–22. (ACS Symposium Series 702.)

Fahey, J.W., Zhang, Y., Talalay, P., (1997). Broccoli sprouts: an exceptionally rich source of inducers of enzymes that protect against chemical carcinogens. *Proc Natl Acad Sci USA*. 94, 10367–10372.

Fahey, J.W., Zalcmann, A.T., Talalay, P. 2001. The chemical diversity and distribution of glucosinolates and isothiocyanates among plants. *Phytochemistry* 56(1), 5-51. [corrigendum: *Phytochemistry* 59: 237].

Favell, DJ., (1998). A comparison of the vitamin C content of fresh and frozen vegetables. *Food Chem.* 62, 59-64.

Fenwick, G.R., Curl, C.L., Griffiths, N.M., Heaney, R.K., Price, K.R., (1990). Bitter principles in food plants. *In*: Rouseff RL, ed. Bitterness in foods and beverages; developments in food science 25. Amsterdam: Elsevier, pp 205–250.

Fernández-León, M.F., Fernández-León, A.M., Lozano, M., Ayuso, M.C., González-Gómez, D., (2013). Altered commercial controlled atmosphere storage conditions for 'Parhenon' broccoli plants (*Brassica oleracea* L. var. *italica*). Influence on the outer quality parameters and on the health-promoting compounds. *LWT - Food Sci Technol* 50(2), 665-672.

Ferreres, F., Valentao, P., Llorach, R., Pinheiro, C., Cardoso, U., Pereira, J.A., Sousa, C., Seabra, R.M., Andrade, P.B., (2005). Phenolic compounds in external leaves of tronchuda cabbage (*Brassica oleracea* L. var. Costata DC). *J. Agric. Food Chem.* 53, 2901- 2907.

Force, L.E., O'Hare, T.J., Wong, L.S., Irving, D.E., (2007). Impact of cold storage on glucosinolate levels in seed-sprouts of broccoli, rocket, white radish and kohl-rabi. *Postharvest Biol Technol* 44(2), 175-178.

Fortier, E., Desjardins, Y., Tremblay, N., Bélec, C., Côté, M., (2010). Influence of irrigation and nitrogen fertilization on broccoli polyphenolics concentrations. *Acta Hort.* 856, 55–62.

Francisco, M., Cartea M.E, Soengas P, Velasco P., (2011). Effect of genotype and environmental conditions on health-promoting compounds in *Brassica rapa*. *J Agric Food Chem.* 59(6), 2421-31.

Francisco, M., Moreno, D.A., Cartea, M.E., Ferreres, F., Garcia-Viguera, C., Velasco, P., (2009). Simultaneous identification of glucosinolates and phenolic compounds in a representative collection of vegetable *Brassica rapa*. *J. Chromatogr. A* 1216, 6611-6619.

Fresco, P., Borges, F., Marques, M.P.M., Diniz, C., (2010). The Anticancer Properties of Dietary Polyphenols and its Relation with Apoptosis. *Curr. Pharm. Des.*16, 114-134.

FSA, (2002). Food Standards Agency. Consumer Attitudes to Food Standards. London: The Stationery Office, 2002.

Furuya, A., Uozaki, M., Yamasaki, H., Arakawa, T., Arita, M., Koyama, A.H., (2008). Antiviral effects of ascorbic and dehydroascorbic acids in vitro. *Int. J. Mol. Med.* 22, 541.

Galizzi, G., Venturini, L., (1996). Economics of innovation: the case of the food industry. Physica-Verlag, Heidelberg, pp. 319-327.

Gawlik-Dziki, U., (2008). Effect of hydrothermal treatment on the antioxidant properties of broccoli (*Brassica oleracea* var. *botrytis italica*) florets. *Food Chem.* 109(2), 393–401.

Gębczyński, P., Lisiewska, Z., (2006). Comparison of the level of selected antioxidative compounds in frozen broccoli produced using traditional and modified methods. *Innov Food Sci Emerg Technol* 7, 239–245

Giamoustaris A., Mithen R., (1996). Genetics of aliphatic glucosinolates.. Side-chain modification in *Brassica oleracea*. *Theor Appl Genet* 93,1006–1010.

Gigolashvili, T., Yatusevich, R., Rollwitz, I., Humphry, M., Gershenzon, J., Flügge, U-I., (2009). The plastidic bile acid transporter 5 is required for the biosynthesis of methionine-derived glucosinolates in *Arabidopsis thaliana*. *Plant Cell.* 21, 1813-1829.

Gil, M.I., Ferreres, E., Tomas-Barberan, F.A., (1999). Effect of postharvest storage and processing on the antioxidant constituents (flavonoids and vitamin C) of fresh cut spinach. *J Agric Food Chem.* 47, 2213–2217

Gil, M.I., Ferreres, F., Tomas-Barberan, F.A., (1998). Effect of modified atmosphere packaging on the flavonoids and vitamin C content of minimally processed Swiss chard (Beta 6ulgaris Subspecies cycla). *J. Agric. Food Chem.* 46, 2007–2012.

Giovannucci, E., Rimm, E.B., Liu, Y., Stampfer, M.J., Willett, W.C., (2003). A prospective study of cruciferous vegetables and prostate cancer. *Cancer Epidemiol. Biomarkers Prev.* 12, 1403–1409.

Gitz, D.C., Liu, L., McClure, J.W., (1998). Phenolic metabolism, growth and UV-B tolerance in phenylalanine ammonia lyase inhibited red cabbage seedlings. *Phytochemistry* 49, 377–386.

Glanz, K., Basil, M., Maibach, E., Goldberg, J., Snyder, D., (1998). Why Americans eat what they do: taste, nutrition, cost, convenience and weight control concerns as influences on food consumption. *J Am Diet Assoc.* 98, 1118–1126.

Glanz, K., Brug, J., van Assema, P., (1997). Are awareness of dietary fat intake and actual fat consumption associated?–a Dutch-American comparison. *Eur J Clin Nutr.* 51(8), 542–547.

Gliszczyńska-Swigło, A., Ciska, E., Pawlak-Lemańska, K., Chmielewski, J., Borkowski, T., Tyrakowska, B., (2006). Changes in the content of health-promoting compounds and antioxidant activity of broccoli after domestic processing. *Food Addit Contam.* 23(11), 1088-1098.

Gökmen, V., Kahraman, N., Demir, N., Acar, J., (2000). Enzymatically validated liquid chromatographic method for the determination of ascorbic and dehydroascorbic acids in fruit and vegetables. *J Chromatogr A* 881, 309-316.

Gomez-Lopez, V., Ragaert, P., Debevere, J., Devlieghere, F., (2008). Decontamination methods to prolong the shelf-life of minimally processed vegetables, state-of-the-art. *Crit. Rev. Food Sci. Nutr.* 48, 487–495.

Gonçalves, E.M., Pinheiro, J., Alegria, C., Abreu, M., Brandão, T.R.S., Silva. C.L.M., (2009). Degradation kinetics of peroxidase enzyme, phenolic content, and physical and sensorial characteristics in broccoli (*Brassica oleracea* L. ssp. *Italica*) during blanching. *J. Agric. Food Chem.* 57(12), 5370–5375

Gonçalves, E.M., Abreu, M., Brandão, T.R.S., Silva, C.L.M., (2011). Degradation kinetics of colour, vitamin C and drip loss in frozen broccoli (*Brassica oleracea* L. ssp. *Italica*) during storage at isothermal and non-isothermal conditions. *Int J Refrig* 34(8), 2136–2144.

Gonzalez-Aguilar, G.A., Celis, J., Sotelo-Mundo, R:R., de la Rosa, L.A., Rodrigo-Garcia, J., et al., (2008). Physiological and biochemical changes of different fresh-cut mango cultivars stored at 5°C. *Int. J. Food Sci. Technol.* 43, 91-101.

Granado-Lorencio, F., Olmedilla-Alonso, B., Herrero-Barbudo, C., Sánchez-Moreno, C., de Ancos, B., Martínez, J.A., Pérez-Sacristán, B., Blanco-Navarro, I., (2008). Modified-atmosphere packaging (MAP) does not affect the bioavailability of tocopherols and carotenoids from broccoli in humans: A cross-over study. *Food Chem.* 106, 1070-1076.

Grene, R., (2002). Oxidative stress and acclimation mechanisms in plants. Arabidopsis Book;1:e0036.

Groff, J.L., Gropper, S.S., Hunt, S.M., (1995). The water soluble vitamins. *In*: Advanced Nutrition and Human Metabolism. West PublishingCompany, Minneapolis, pp. 222–231.

Grubb, C.D., Abel, S. (2006). Glucosinolate metabolism and its control. *Trends Plant Sci.* 11, 89–100.

Hagen, S.F., Borge, G.I.A., Solhaug, K.A., Bengtsson, G.B., (2009). Effect of cold storage and harvest date on bioactive compounds in curly kale (*Brassica oleracea* L. var. *acephala*). *Postharvest Biol Technol* 51(1), 36–42.

Halkier, B.A., Du, L., (1997). The biosynthesis of glucosinolates. *Trends Plant Sci* 2, 425–431.

Halvorsen, B.L., Carlsen, M.H., Phillips, K.M., Bohn, S.K., Holte, K., Jacobs, D.R., (2006). Content of redox-active compounds (i.e., antioxidants) in foods consumed in the United States. *Am J Clin Nutr* 84(1), 95–135.

Hansen, M., Moller, P., Sorensen, H., Detrejo, M.C., (1995). Glucosinolates in broccoli stored under controlled-atmosphere. *J. Am. Soc. Hortic. Sci.* 120, 1069–1074

Harbaum-Piayda, B., Walter, B., Bengtsson, G.B., Hubbermann, E.M., Bilger, W., Schwarz, K., (2010). Influence of pre-harvest UV-B irradiation and normal or controlled atmosphere storage on flavonoid and hydroxycinnamic acid contents of pak choi (*Brassica campestris* L. ssp. *chinensis* var. *communis*). *Postharvest Biol Technol* 56(3), 202-208.

Harris, J., Felix, L., Miners, A., Murray, E., Michie, S., Ferguson, E., Free, C., Lock, K., Landon, J., Edwards, P., (2011). Adaptive e-learning to improve dietary behaviour: a systematic review and cost-effectiveness analysis. *Health Technol Assess.* 15(37), 1-160.

Hart, D.J., Scott, K.J., (1995) Development and evaluation of an HPLC method for the analysis of carotenoids in foods, and the measurement of the carotenoid content of vegetables and fruits commonly consumed in the UK. *Food Chem* 54,101-111

Heimler, D., Vignolini, P., Dini, M.G., Vincieri, F.F., Romani, A., (2006). Antiradical activity and polyphenol composition of local Brassicaceae edible varieties. *Food Chem.* 99, 464–469.

Heinonen, S., Nurmi, T., Liukkonen, K., Poutanen, K., Wahala, K., Deyamat, T., Nishibe, S., Aldercreuts, H., (2001). *In vitro* metabolism of plant lignans: New precursors of mammalian lignans entrolactone and enterodiol. *J Agric Food Chem* 49, 3178-3186.

Heinonen, M.I., Ollilainen, V., Linkola, E.K., Varo, P.T., Koivistoinen, P:E., (1989). Carotenoids in Finnish foods: vegetables, fruits, and berries. *J Agric Food Chem* 37, 655-659

Hill, C.B., Williams, P.H., Carlson, D.G., Tookey, H.L., (1987). Variation in glucosinolates in oriental brassica vegetables. *J. Am. Soc. Hortic. Sci.* 122, 309-313.

Hodges, M., Munro, K.D., Forney, C.F., McRae, K.B., (2006). Glucosinolate and free sugar content in cauliflower (*Brassica oleracea* var. *botrytis* cv. Freemont) during controlled-atmosphere storage. *Postharvest Biol Technol* 40(2), 123-132D.

Holick, C.N., Michaud, D.S., Stolzenberg-Solomon, R., Mayne, S.T., Pietinen, P., Taylor, P.R., Virtamo, J., Albanes, D., (2002). Dietary carotenoids, serum beta-carotene, and retinol and risk of lung cancer in the alpha-tocopherol, beta-carotene cohort study. *Am J Epidemiol.* 156(6), 536–547.

Hollósy, F., (2002). Effects of ultraviolet radiation on plant cells. *Micron* 33, 179–197.

Howard, L.A., Wong, A.D., Perry, A.K., Klein, B.P., (1999). B-carotene and ascorbic acid retention in fresh and processed vegetables. *J. Food Sci.* 64, 929–936.

Howard, L.R., Hernandez-Brenes, C., (1998). Antioxidant content and market quality of jalapeno rings as affected by minimal processing and modified atmosphere packaging. *J. Food Qual.* 21, 317-327.

Hsu, F.C., Wirtz, M., Heppel, S.C., Bogs, J., Krämer, U., Khan, M.S., Bub, A., Hell, R., Rausch, T., (2011). Generation of Se-fortified broccoli as functional food: impact of Se fertilization on S metabolism. *Plant Cell Environ*. 34(2), 192-207.

Hunter K.J., Fletcher, J.M. (2002). The antioxidant activity and composition of fresh, frozen, jarred and canned vegetables. *Innov Food Sci Emer Technol.* 3, 399-406.

Hwang, Eun-Sun. Jang, Mi-Ran, Kim, Gun-Hee., (2012). Effects of storage condition on the bioactive compound contents of Korean cabbage. *Food Sci. Biotechnol.* 21(6), 1655-1661.

IFPA, 2009. Food Safety Guidelines for the Fresh-Cut Produce Industry, J.R. Gorny, Ed., 4th edn. IFPA, Alexandria, VA.

Iglesias-Acosta, M., Martínez-Ballesta, M.C., Teruel, J.A., Carvajal, M., (2010). The response of broccoli plants to high temperature and possible role of root aquaporins. *Environ. Exp. Bot.* 68(1), 83–90.

Ishii, G., Saijo, R., (1987). Effect of season, soil type, sulfate level, mulching and plant density on isothiocyanates content in radish root juice (*Raphanus sativus* L.). *J. Japan. Soc. Hort. Sci.* 56, 313–320.

Jacoba, R.A., 1999. Vitamin C. *In*: Modern Nutrition in Health and Disease, 9th. Williams & Wilkins, Baltimore, pp. 467–473.

Jacobsson, A., Nielsen, T., Sjöholm, I., Wendin, K., (2004). Influence of packaging material and storage condition on the sensory quality of broccoli. *Food Qual Prefer* 15(4), 301-310.

Jahangir, M., Kim, H.K., Choi, Y.H., Verpoorte, R., (2009a). Health-affecting compounds in Brassiceae. *Comp. Rev. Food Sci. Food Safety*. 8, 1-43.

Jahangir, M., Abdel-Farid, I.B., Kim, H.K., Choi, Y.H., Verpoorte, R., (2009b). Healthy and unhealthy plants: the effect of stress on the metabolism of Brassicaceae. *Environ. Exp. Bot.* 67, 23–33.

Jansen, M.A.K., Gaba, V., Greenberg, B.M., (1998). Higher plants and UV-B radiation: balancing damage, repair and acclimation. *Trends Plant Sci.* 3, 131–135.

Jansen, M.A.K., Hectors, K, O'Brien, N.M., Guisez, Y., Potters, G., (2008). Plant stress and human health: do human consumers benefit from UV-B acclimated crops? *Plant Sci.* 175, 449–458.

Jansen, M.A.K., van den Noort, R.E., Adillah Tan, M.Y., Prinsen, E., Lagrimini, L.M., Thorneley, R.N.F., (2001). Phenol-Oxidizing Peroxidases Contribute to the Protection of Plants from Ultraviolet Radiation Stress. *Plant Physiol.* 126, 1012–1023.

Jariwalla, R. J., Harakeh, S., (1996). Antiviral and immunomodulatory activities of ascorbic acid. *Subcell. Biochem.* 25, 213-231.

Jaworska G., Kmiecik, W., Maciejaszek, I., (2001). Comparasion of the quality of canned spinach (*Spinacia Oleracea* L.) and New Zealand spinach (*Tetragonoa Expansa* Murr.), *EJPAU* 4(2), 04.

Jones, R.B., Faragher, J.D., Winkler, S., (2006). A review of the influence of postharvest treatments on quality and glucosinolate content in broccoli (*Brassica oleracea* var. *italica*) heads. *Postharvest Biol Technol.*, 41, 1–8.

Kaur, S., Gupta, S.K., Sukhija, P.S., Munshi, S.K., (1990). Accumulation of glucosinolates in developing mustard (Brassica juncea L.) seeds in response to sulfur application. *Plant Sci.* 66, 181–184.

Kerr, D.A., Pollard, C.M., Howat, P., Delp, E.J., Pickering, M., Kerr, K.R., Dhaliwal, S.S., Pratt, I.S., Wright, J., Boushey, C.J., (2012). Connecting Health and Technology (CHAT): protocol of a randomized controlled trial to improve nutrition behaviours using mobile devices and tailored text messaging in young adults. *Health* 12, 477-487.

Keum, Y.-S., Jeong, W.-S., Kong, A.N.T., (2004). Chemoprevention by isothiocyanates and their underlying molecular signaling mechanisms. *Mutat. Res.* 555, 191–202.

Khan, M.A.M., Ulrichs, C., Mewis, I., (2010). Influence of water stress on the glucosinolate profile of *Brassica oleracea* var. *italica* and the performance of *Brevicoryne brassicae* and *Myzus persicae*. *Entomol. Exp. Appl.* 137(3), 229–236.

Khan, M.A.M., Ulrichs, C., Mewis, I., (2011). Water stress alters aphid induced glucosinolate response inBrassica oleracea var. italica differently. *Chemoecology* 21, 235–242.

Ki-Hwan, S., Nack-Kie, S., kap-Suk, K., Cheol-Woo, A., Know-Il, S., (1992). Analysis of glucosinolates and change of contents during processing and storage in cruciferous vegetables. *J Korean Soc. Food Nutr.* 21, 43-48.

Kim, S.J., Matsuo, T., Watanabe, M., Watanabe, Y., (2002). Effect of nitrogen and sulphur application on the glucosinolate content in vegetable turnip rape (*Brassica rapa* L.). *Soil Sci. Plant Nutr.* 48, 43–49.

Kjær, A., (1976). Glucosinolates in the Cruciferae. *In*: J.G. Vaughan, A.J. MacLeod, B.M.G. Jones (Eds.), The Biology and Chemistry of the Cruciferae, Academic Press, London (1976), pp. 207–219

Kopsell, D.E., Kopsell, D.A., Randle, W.A., Coolong, T.W., Sams, C.E., Curran-Celentano, J., (2003). Kale carotenoids remain stable while flavor compounds respond to changes in sulfur fertility. *J. Agric. Food Chem.* 51, 5319–5325.

Koritsas, V.M., Lewis, J.A, Fenwick, G.R., (1991). Glucosinolate responses of oil seed rape, mustard and kale to mechanical wounding and infestation by cabbage stem flea beetle (*Psylliodes chuysocephala*). *Am Appl Biol* 118, 209-221.

Korus, A., (2011). Effect of preliminary processing, method of drying and storage temperature on the level of antioxidants in kale (*Brassica oleracea* L. var. *acephala*) leaves. *LWT - Food Sci. Technol* 44(8), 1711-1716.

Korus, A., Lisiewska. Z., (2011). Effect of preliminary processing and method of preservation on the content of selected antioxidative compounds in kale(*Brassica oleracea* L. var. i) leaves. *Food Chem* 129, 149-154.

Koukol, J., Conn, E.E., (1961). The metabolism of aromatic compounds in higher plants. *J Biol Chem* 236, 2692–2697.

Krieger-Liszkay, A., Fufezan, C., Trebst, A., (2008). Singlet oxygen production in photosystem II and related protection mechanism. *Photosynth Res* 98, 551–564

Kristal, A.R., Patterson, R.E., Glanz, K., Heimendinger, J., Hebert, J.R., Feng, Z., Probart, C., (1995). Psychosocial correlates of healthful diets: baseline results from the working well study. *Prev Med* 24(3), 221-228.

Krizek, D.T., Britz, S.J., Mirecki, R.M., (1998). Inhibitory effects of ambient levels of solar UV-A and UV-B radiation on growth of cv. New Red Fire lettuce. *Physiol. Plant.* 103, 1–7.

Krizek, D.T., Mirecki, R.M., Britz, S.J., (1997). Inhibitory effects of ambient levels of solar UV-A and UV-B radiation on growth of cucumber. *Physiol. Plant.* 100, 886–893.

Kuhlmann, F., Müller, C., (2009). Development-dependent effects of UV radiation exposure on broccoli plants and interactions with herbivorous insects. *Environ. Exp. Bot.* 66, 61–68.

Kurilich, A.C., Tsau, J.C., Brown, A., Howard, L., Klein, B.P., Jeffery, E.H., Kushad, M., Walling, M.A., Juvik, A., (1999). Carotene, tocopherol and ascorbate contents in subspecies of *Brassica oleracea*. *J. Agric. Food Chem.* 47, 1576-1581.

Kushad, M.M., Brown, A.F., Kurilich, A:C., Juvik, J.A., Klein, B.P., Wallig, M.A., Jeffery, E.H., (1999). Variation of Glucosinolates in Vegetable Crops of *Brassica oleracea*. *J. Agric. Food Chem.* 47, 1541-1548.

Lechner, L., Brug, J., Vries, H.D., (1997). Misconceptions of fruit and vegetable consumption: differences between objective and subjective estimation of intake. *J Nutr Educ* 29(6), 313-320.

Lee, S.K., Kader, A.A., (2000). Preharvest and postharvet factors influencing vitamin C content of horticultural crops. *Postharvest Biol Technol* 20(3), 207–220.

Lefsrud, M.G. Kopsell, D.A., Sams, C.E., (2008). Irradiance from Distinct Wavelength Light-emitting Diodes Affect Secondary Metabolites in Kale. *HortScience* 43(7), 2243-2244.

Leja, M., Starzynska, A., Mareczek, M., Rozek, S., (2001). Antioxidant ability of broccoli flower buds during short-term storage. *Food Chem.* 72, 219-222.

Lin, C.H, Chang, C.Y., (2005). Textural change and antioxidant properties of broccoli under different cooking treatments. *Food Chem.* 90(1-2), 9–15.

Lisiewska, Z., Kmiecik, W., (1996). Effects of level of nitrogen fertilizer, processing conditions and period of storage of frozen broccoli and cauliflower on vitamin C retention. *Food Chem.* 57(2), 267–270.

Liu, R.H., (2004). Potential synergy of phytochemicals in cancer prevention: mechanism of action. *J Nutr.* 134(12 Suppl), 3479S-3485S.

Llorach, R., Espín, J.C., Tomás-Barberán, F.A., Ferreres, F., (2003). Valorization of cauliflower (*Brassica oleracea* L. var *Botrytis*) by-products as a source of antioxidant phenolics. *J. Agric. Food Chem.* 51, 2181-2187

Lo Scalzo, R., Iannoccari, T., Genna, A., Di Cesare, L. F., Viscardi, D., Ferrari, V., et al., (2008). Organic vs. conventional field trials: The effect on cauliflower quality. *In* Proceedings of the 16th IFOAM organic world congress, cultivate the future based on science, 2nd conference of the international society of organic agriculture research ISOFAR, Modena, Italy, ID code 11758.

Lombardi-Boccia, G., Lucarini, M., Lanzi, S., Aguzzi, A., Cappelloni, M., (2004). Nutrients and antioxidant molecules in yellow plums (*Prunus domestica* L.) from conventional and organic productions: A comparative study. *J. Agric. Food Chem.* 52, 90–94.

López-Berenguer, C., García-Viguera, C., Carvajal, M., (2006). Are root hydraulic conductivity responses to salinity controlled by aquaporins in broccoli plants? *Plant Soil* 279, 13–23.

López-Berenguer, C., Carvajal, M., Moreno, D.A., Garcia-Viguera, C., (2007). Effects of microwave cooking conditions on bioactive compounds present in broccoli inflorescences. *J Agric Food Chem.* 55(24), 10001–10007.

López-Berenguer, C., Martínez-Ballesta, M.C., García-Viguera, C., Carvajal, M., (2008). Leaf water balance mediated by aquaporins under salt stress and associated glucosinolate synthesis in broccoli. *Plant Sci* 174, 321–328.

Mackerness, S., (2000). Plant responses to ultraviolet-B (UV-B:280–320 nm) stress: What are the key regulators? *Plant Growth Regul* 37, 27-39.

Martinez, J.A., De Ancos, B., Sánchez-Moreno, C., Cano, M.P., (2007). Protective effect of minimally processing on bioactive compounds of broccoli (*Brassica oleracea* L. var *italica*) during refrigerated storage and after cold chain rupture. *In*: Abstracts of V Iberoamerican Congress in Postharvest and Agricultural Exports. Postharvest Group of Polytechnic University of Cartagena and Iberoamerican Postharvest Association (Eds.), pp 694-706, Cartagena, Murcia, Spain.

Martínez, S., Pérez, N., Carballo, J., Franco, I., (2013). Effect of blanching methods and frozen storage on some quality parameters of turnip greens ("grelos"). *LWT - Food Sci. Technol.* 51(1), 383–392.

Martínez-Ballesta, M.C., López-Pérez, L., Hernández, M., López-Berenguer, C., Fernández-García, N., Carvajal, M., (2008). Agricultural practices for enhanced human health. *Phytochem Rev.* 7(2), 251-260.

Martínez-Hernández, G.B., Artés-Hernández, F., Gómez, P.A., Artés, F., (2013). Quality changes after vacuum-based and conventional industrial cooking of kailan-hybrid broccoli throughout retail cold storage. *LWT - Food Sci. Technol* 50(2), 707–714

McKenzie, R.L., Aucamp, P.J., Bais, A.F., Björn, L.O., Ilyas, M., (2007). Changes in biologically-active ultraviolet radiation reaching the earth's surface. *Photochem. Photobiol.* 6, 218–231.

Mewis, I., Schreiner, M., Nguyen, C.N., Krumbein, A., Ulrichs, C., Lohse, M., Zrenner, R., (2012). UV-B Irradiation Changes Specifically the Secondary Metabolite Profile in Broccoli Sprouts: Induced Signaling Overlaps with Defense Response to Biotic Stressors. *Plant Cell Physiol.* 53(9), 1546–1560.

Miglio, C., Chiavaro, E., Visconti, A., Fogliano, V., Pellegrini, N., (2008). Effects of different cooking methods on nutritional and physicochemical characteristics of selected vegetables. *J Agric Food Chem.* 56(1),139–147.

Milder, I.E., Arts, I.C., van de Putte, B., Venema, D.P., Hollman, P.C., (2005). Lignan contents of Dutch plant foods: a database including lariciresinol, pinoresinol, secoisolariciresinol and matairesinol. *Br J Nutr.* 93, 393–402

Mithen, R., Faulkner, K., Magrath, R., Rose, P., Williamson, G., Marquez, J., (2003). Development of isothiocyanate-enriched broccoli and its enhanced ability to induce phase 2 detoxification enzymes in mammalian cells. *Theor Appl Gen* 106, 727–734.

Mithen, R.F., Dekker, M., Verkerk, R., Rabot, S., Johnson, I.T., (2000), The nutritional significance, biosynthesis and bioavailability of glucosinolates in human foods. *J Sci Food Agri.* 80, 967-984.

Mittler, R., (2002). Oxidative stress, antioxidants and stress tolerance. *Trends Plant Sci.* 7(9), 405-10.

Mondragón-Portocarrero, A., Pena-Martinez, B., Fernandez-Fernandez, E., Romero-Rodriguez, A., Vazquez-Oderiz, L., (2006). Effects of different pre-freezing blanching procedures on the physicochemical properties of *Brassica rapa* leaves (Turnip Greens, Grelos). *Int. J. Food Sci. Technol.* 41, 1067-72.

Moreno, D.A., Carvajal, M., López-Berenguer, C., García-Viguera, C., (2006). Chemical and biological characterisation of nutraceutical compounds of broccoli. *J. Pharm. Biomed. Anal.* 41(5), 1508-22.

Moreno, D.A., Pérez-Balibrea, S., Gil-Izquierdo, A., García-Viguera, C., (2010). Acylated anthocyanins in broccoli sprouts. *Food Chem.* 123, 358-363.

Muller, H. (1997). Determination of the carotenoid content in selected vegetables and fruit by HPLC and photodiode array detection. *Zeitschrift fur Lebensmittel- Untersuchung und Forschung A.* 204, 88–94.

Murcia, M. A., Lopez-Ayerra, B., Martinez-Tome, M., Vera, A. M., Garcia-Carmona, F. (2000). Evolution of ascorbic acid and peroxidase during industrial processing of broccoli. *J Sci Food Agric* 80, 1882–1886.

Murcia, M.A., Martínez-Tomé, M., Nicolás, M.C., Vera, A.M., (2003). Extending the shelf-life and proximate composition stability of ready to eat foods in vacuum or modified atmosphere packaging. *Food Microbiol* 20, 671-679.

Nath, A., Bagchi, B., Misra, L.K., Bidyut, C.D., (2011). Changes in post-harvest phytochemical qualities of broccoli florets during ambient and refrigerated storage. *Food Chem.* 127, 1510-1514.

Nepal, V.P., Mgbere, O., Banerjee, D., Arafat, R.R., (2011). Disparities in Fruits and Vegetables Consumption in Houston, Texas: Implications for Health Promotion. *Journal of Primary Care & Community Health* 2(3),142-147.

Nielsen, J.K., Olsen, C.E., Petersen, M.K., (1993). Acylated flavonol glycosides from cabbage leaves. *Phytochemistry* 49, 2171–2176.

Nilsson, J., Olsson, K., Engqvist, G., Ekvall, J., Olsson, M., Nyman, M., et al., (2006). Variation in the content of glucosinolates, hydroxycinnamic acids, carotenoids, total antioxidant capacity and low-molecularweight carbohydrates in *Brassica* vegetables. *J Sci Food Agric* 86, 528–538.

Ninfali, P., Bacchiocca, M., (2003). Polyphenols and antioxidant capacity of vegetables under fresh and frozen conditions. *J Agric Food Chem.* 51(8), 2222-2226.

Oerlemans, K., Barrett, D.M., Bosch Suades, C., Verkerk, R., Dekker, M., (2006): Thermal degradation of glucosinolates in red cabbage. *Food Chem.* 95, 19–29.

Oh, M.-M., Rajashekar, C.B. (2009). Antioxidant content of edible sprouts: effects of environmental shocks. *J Sci Food Agric* 89(13), 2221-2227.

Olivera, D.F., Viňa, S.Z., Marani, C.M., Ferreyra, R.M., Mugridge, A., Chaves, A.R., Mascheroni, R.H., (2008). Effect of blanching on the quality of Brussel sprouts (*Brassica oleracea* L. *gemmifera DC*) after frozen storage. *J. Food Eng.* 84, 148-155.

Oliviero, T., Verkerk, R., Dekker, M., (2012). Effect of water content and temperature on glucosinolate degradation kinetics in broccoli (*Brassica oleracea* var. *italica*). *Food Chem.* 132(4), 2037–2045.

Omenn, G.S., Goodman, G.E., Thornquist, M.D., Balmes, J., Cullen, M.R., Glass, A., Keogh, J.P., Meyskens, F.L., Valanis, B., Williams, J.H., Barnhart, S., Hammar, S., (1996). Effects of a combination of beta carotene and vitamin A on lung cancer and cardiovascular disease. *N Engl J Med.* 334(18), 1150-1155.

Padilla, G., Cartea, M.E., Velasco, P., de Haro, A., Ordás, A., (2007). Variation of glucosinolates in vegetable crops of *Brassica rapa*. *Phytochemistry* 68, 536–545.

Paolini, M., (1998). Brussels sprouts: An exceptionally rich source of ambiguity for anticancer strategies. *Toxicol. Appl. Pharmacol.* 152, 293–294.

Parkinson, R. W. U.S. Patent 4 424 232, 1984.

Paschold, P., Kleber, J., Adam, S., Bognar, A., Tauscher, B., (2000). Einfluss von Bewässerung und N-Düngung auf Ertragund Sulforaphangehalt von Brokkoli (*Brassica oleracea*). *In:* Proceedings of the 35th conference of the German Society of Quality Research, Karlsruhe, pp 57–66.

Pereira, D.M., Valentao, P., Pereira, J.A., Andrade, P.B., (2009). Phenolics: From Chemistry to Biology. *Molecules* 14, 2202-2211.

Pereira, F.M.V., Rosa, E., Fahey, J.W., Stephenson, K.K., Carvalho, R., Aires, A., (2002) Influence of temperature and ontogeny on the levels of glucosinolates in broccoli (*Brassica oleracea* var. *italica*) sprouts and their effect on the induction of mammalian phase 2 enzymes. *J Agric Food Chem.* 50, 6239–6244.

Perez, C.E., (2002). Fruit and vegetable consumption. *Health Rep* 13, 23-31.

Pérez-Balibrea, S., Moreno, D.A., Gracia-Viguera, C., (2008). Influence of light on health promoting phytochemicals of broccoli sprouts. *J. Sci. Food Agric.* 88, 904-910.

Picchi, V., Miglioria, C., Lo Scalzo, R., Campanelli, G., Ferrari, V., Di Cesare, L.F., (2012). Phytochemical content in organic and conventionally grown Italian cauliflower. *Food Chem.* 130(3), 501–509.

Piironen, V., Syvaoja, E.-L., Varo, P., Salminen, K., & Koivistoinen, P., (1986). Tocopherols and tocotrienols in Finnish foods: Vegetables, fruits, and berries. *J Agric Food Chem.* 34, 742–746.

Pinheiro-Sant'ana, H.M., Guinazi, M., Oliveira, D.S., Della Lucia, C.M., Reis, B.L., Brandão, S.C., (2011). Method for simultaneous analysis of eight vitamin E isomers in various foods by high performance liquid chromatography and fluorescence detection. *J Chromatogr A.* 1218(47):8496-502.

Plumb, G.W., Lambert, N., Chambers, S.J., Wanigatunga, S., Heaney, R.K., Plumb, J.A., et al., (1996). Are whole extracts and purified glucosinolates from cruciferous vegetables antioxidants? *Free Radical Res.* 25(1), 75–86.

Plumb, G.W., Price, K.R., Rhodes, M.J.C., Williamson, G., (1997). Antioxidant properties of the major polyphenolic compounds in broccoli. *Free Radical Res.* 27, 429-435.

Podsedek, A., (2007). Natural antioxidants and antioxidant capacity of *Brassica* vegetables: A review. *LWT-Food Sci. Technol.* 40(2), 1–11.

Podsedek, A., Sosnowska, D., Redzynia, M., Anders, B., (2006). Antioxidant capacity and content of *Brassica oleracea* dietary antioxidants. *Int. J. Food Sci. Technol.* 41, 49-58.

Podsedek, A., Wilska-Jeska, J., Anders, B., Markowski, J., (2000). Compositional characterization of some apple varieties. *Eur Food Res Technol* 210(4), 268-272.

Povey, R., Conner, M., Sparks, P., James, R., Shepherd, R., (1998). Interpretations of healthy and unhealthy eating, and implications for dietary change. *Health Educ. Res.* 13, 171–83.

Pramanik, B.K., Matsui, T., Suzuki, H., Kosugi, Y., (2004). Changes in activities of sucrose synthase and sucrose phosphate synthase and sugar contente during postharvest senescence in two broccoli cultivars. *Asian J. Plant Sci.* 3, 398-402.

Puupponen-Pimia, R., Hakkinen, S. T., Aarni, M., Suortti, T., Lampi, A. M., Eurola, M., et al. (2003). Blanching and long-term freezing affect various bioactive compounds of vegetables in different ways. *J. Sci. Food Agric.*, 83, 1389–1402.

Quinsac, A., Charrier, A., Ribaillier, D., Peron, J.Y., (1994). Glucosinolates in etiolated sprouts of sea-kale (Crambe-Maritima L). *J. Sci. Food Agric.*, 65, 201–207.

Raffo, A., Baiamonte, I., Paoletti, F., (2008). Changes in antioxidant and taste-related compounds contend during cold storage of fresh-cut red sweet peppers. *Eur. Food Res. Technol.* 226, 1167-1174.

Rangkadilok, N., Nicolas, M.E., Bennett, R.N., Eagling, D.R., Premier, R.R., Taylor, P.J., (2004). The effect of sulphur fertilizer on glucoraphanin levels of broccoli (B. oleracea L. var. italica) at different growth stages. *J. Agric. Food Chem.* 52, 2632-2639.

Rangkadilok, N., Nicolas, M.E., Bennett, R.N., Premier, R.R., Eagling, D.R., Taylor, P.W.J., (2002). Developmental changes of sinigrin and glucoraphanin in three *Brassica* species (*Brassica nigra*, *Brassica juncea* and *Brassica oleracea* var. *italica*). *Sci Hort* 96, 11-26.

Rasmussen, M., Krølner, R., Klepp, K.I., Lytle, L., Brug, J., Bere, E., Due, P., (2006). Determinants of fruit and vegetable consumption among children and adolescents: a review of the literature. Part I: quantitative studies. *Int J Behav Nutr Phys Act* 3, 22.

Reed, d.W., Tukey, J., (1982). Light intensity and temperature effects on epicuticular wax morphology and internal cuticule ultrastructure of carnation and Brussels sprouts leaf cuticules. *J. Amer. Soc. Hort. Sci.* 109, 417-420.

Reime, B., Novak, P., Born, J., Hagel, E., Wanek, V., (2000). Eating habits, health status, and concern about health: a study among 1641 employees in the German metal industry. *Prev Med* 30(4), 295-301

Reyes, L.F., Villarreal, J.E., Cisneros-Zevallos, L., (2007). The increase in antioxidant capacity after wounding depends on the type of fruit or vegetable tissue. *Food Chem.* 101, 1254–1262.

Ricciuto, L., Tarasuk, V., Yatchew, A., (2006). Socio-demographic influences on food purchasing among Canadian households. *Eur J Clin Nutr* 60, 778-790.

Rice-Evans, C.A, Miller, N.J, Bolwellm P.G, Bramley, P.M, Pridham, J.B., (1995). The relative antioxidant activities of plant-derived polyphenolic flavonoids. *Free Radical Res.* 22(4), 375-83.

Rice-Evans, C.A, Miller, N.J., Paganga, G., (1996). Structure-antioxidant activity relationships of flavonoids and phenolic acids. *Free Radical Biol. Med.*, 20(7), 933-956. *Erratum in*: Free Radical Biol. Med., 1996, 21(3), 417.

Richards, A., Wijesundera, C., Salisbury, P., (2008). Genotype and growing environment effects on the tocopherols and fatty acids of *Brassica napus* and *B. juncea*. *J. Am. Oil Chem. Soc.* 85, 159–168.

Riediger, N.D., Moghadasian, M.H., (2008). Patterns of Fruit and Vegetable Consumption and the Influence of Sex, Age and Socio-Demographic Factors among Canadian Elderly. *J Am Coll Nutr* 27(2), 306-313.

Robards, K., Prenzler, P. D., Tucker, G., Swatsitang, P., Glover, W., (1999). Phenolic compounds and their role in oxidative processes in fruits. *Food Chem.* 66, 401–436.

Robbins, R.J., Keck, A.S., Banuelos, G., Finley, J.W., (2005). Cultivation conditions and selenium fertilization alter the phenolic profile, glucosinolate, and sulforaphane content of broccoli. *J Med Food.* 8(2), 204-14.

Robinson-O'Brien, R., Story, M., Heim, S., (2009). Impact of Garden-Based Youth *Nutrition* Intervention Programs: A Review. *J Am Diet Assoc.* 109, 273-280.

Radovich, T.J.K., Kleinhenz, M.D., Streeter, J.G., (2005). Irrigation Timing Relative to Head Development Influences Yield Components, Sugar Levels, and Glucosinolate Concentrations in Cabbage. *J. Am. Soc. Hortic. Sci.* 130, 943-949.

Rodrigues, A.S., Rosa, E.A.S., (1999). Effect of post-harvest treatments on the level of glucosinolates in broccoli. *J. Sci. Food Agric.* 79, 1028–1032.

Rooij, J.F.M., (2000). The consumer of the 21st century. *In* A. Boekestein, P. Diederen, W. M. F. Jongen, R. Rabbinge and H. Rutten (Eds.) Towards an agenda for agricultural research in Europe. Wageningen: Pers.

Rosa, E.A.S., (1992). Glucosinolates in cabbage. A study of their variation throughout the growing season. Thesis, UTAD, Vila Real, Portugal.

Rosa, E.A.S., (1997). Daily variation in glucosinolate concentrations in the leaves and roots of cabbage seedling in two constant temperature regimes. *J. Sci. Food Agric.* 73, 364-368.

Rosa, E.A.S., Heaney, R.K., (1993). The effect of cooking and processing on the glucosinolate content – studies on 4 varieties of Portuguese cabbage and hybrid white cabbage. *J. Sci. Food Agric.* 62, 259–265.

Rosa, E.A.S., Heaney, R.K., Fenwick, G.R., Portas, C.A.M., (1997). Glucosinolates in crop plants. *In*: Janick, J. (Ed.), Horticultural Reviws, Jonh Wiley & Sons, New York.

Rosa, E.A.S., Rodrigues, A.S., (1998). The effect of light and temperature on glucosinolate concentration in the leaves and roots of cabbage seedlings. *J. Sci. Food Agric.* 78, 208-212.

Rosa, E.A.S., Rodrigues, A.S., (2001). Total and individual glucosinolate content in 11 broccoli cultivars grown in early and late seasons. *HortScience* 36,56-59.

Roy, M.K., Juneja, L.R, Isobe, S., Tsushida, T., (2009). Steam processed broccoli (*Brassica oleracea*) has higher antioxidant activity in chemical and cellular assay systems. *Food Chem.* 114, 263-269.

Rozema, J., Van de Staaij, J., Björn, L.O., Caldwell, M.M., (1997). UV-B as an environmental factor in plant life: stress and regulation. *Trends Ecol. Evol.* 12, 22-28.

Rungapamestry, V., Duncan, A.J., Fuller, Z., Ratcliffe, B., (2006). Changes in glucosinolate concentrations, myrosinase activity, and production of metabolites of glucosinolates in cabbage (*Brassica oleracea* var. *capitata*) cooked for different durations. *J Agric Food Chem.* 54, 7628–7634.

Rungapamestry, V., Duncan, A.J., Fuller, Z., Ratcliffe, B., (2007). Effect of cooking brassica vegetables on the subsequent hydrolysis and metabolic fate of glucosinolates. *Proc Nutr Soc.* 66, 69–81.

Sakalauskaitė, J., Viškelis, P., Duchovskis, P., Dambrauskienė, E., Sakalauskienė, S., Samuolienė, G., Brazaitytė, A., (2012). UV-B irradiation effects on basil (Ocimum basilicum L.) growth and phytochemical properties. *J. Food Agr. Environ.* 10(3&4), 342-346.

Samec, D., Piljac-Zagarac, J., Bogovic, M., Habjanic, K., Gruz, J. (2012). Antioxidant potency of white (*Brassica oleracea* L. var. *capitata*) and Chinese (*Brassica rapa* L. var. *pekinensis* (Lour.)) cabbage: The influence of development stage, cultivar choice and seed selection. *Sci Hort* 128, 78-83.

Sang, J.P., Bluett, C.A:, Elliott, B.R., Truscott, J.W., (1986). Effect of time of sowing on oil content, erucic acid and glucosinolate contents in rapeseed (*Brassica napus* L. cv. *Marnoo*). *Austral. J. Exp. Agr.* 26, 607-611.

Sangtarash, M.H., Qaderi, M.M., Chinnappa, C.C., Reid, D.M., (2009). Differential responses of two Stellaria longipes ecotypes to ultraviolet-B radiation and drought stress. *Flora-Morphology, Distribution, Functional Ecology of Plants*, 204, 593–603.

Scheerens, J., (2001). Phytochemicals and the consumer: Factors affecting fruit and vegetable consumption and potential for increasing small fruit in the diet. *HortTechnol* 11(4), 547–556.

Schneider, C., (2005). Chemistry and biology of vitamin E. *Mol. Nutr. Food Res.* 49, 7–30.

Schonhof, I., Kläring, H.P., Krumbein, A., Schreiner, M., (2007a). Interaction between atmospheric CO2 and glucosinolates in Broccoli. *J Chem Ecol.* 33(1), 105–114.

Schonhof, I., Kläring, H.-P., Krumbein, A., Claußen, W., Schreiner, M., (2007b). Effect of temperature increase under low radiation conditions on phytochemicals and ascorbic acid in greenhouse grown broccoli. *Agric., Ecosyst. Environ.* 119(1–2), 103–111.

Schouten, R.E., Zhang, X., Verschoor, J.A., Otma, E.C., Tijskens, L.M.M., van Kooten, O., (2009). Development of colour of broccoli heads as affected by controlled atmosphere storage and temperature. *Postharvest Biol Technol.* 51(1), 27-35.

Schreiner, M., Mewis, I., Huyskens-Keil, S., Jansen, M.A.K., Zrenner, R., Winkler, J.B., et al., (2012). UV-B induced secondary plant metabolites—potential benefits for plant and human health. *Crit. Rev. Plant Sci.* 31,229–240.

Schreiner, M., Peters, P., Krumbein, A., (2007). Changes of glucosinolates in mixed fresh-cut broccoli and cauliflower florest in modified atmosphere packaging. *J Food Sci* 72, 585–589.

Serrano, M., Martinez-Romero, D., Guillen, F., Castillo, S., Valero, D., (2006). Maintenance of broccoli quality and functional properties during cold storage as affected by modified atmosphere packaging. *Postharvest Biol Technol.* 39(1), 61–68.

Shapiro, T.A., Fahey, J.W., Wade, K.L., Stephenson, K.K., Talalay, P., (2001). Chemoprotective glucosinolates and isothiocyanates of broccoli sprouts: metabolism and excretion in humans. *Cancer Epidemiol Biomarkers Prev.* 10, 501-508.

Shattuck, V.I., Kakuda, Y., Shelp, B.J., (1991). Effect of low temperature on the sugar and glucosinolate content of rutabaga, *Sci Hort* 48(1–2), 9-19.

Shattuck, V.I., Kakuda, Y., Shelp, B.J., Kakuda, N., (1990). Effects of low-temperature on the quality of turnip roots. *HortScience* 25, 1127.

Shklar, G., Schwartz, J., (1993). Oral cancer inhibition by micronutrients. The experimental basis for clinical trials. *Eur J Cancer B Oral Oncol.* 29B(1), 9-16.

Siemens, D.H., Garner, S.H., Mitchell-Olds, T., Callaway, R.M., (2002). Cost of defense in the context of plant competition: *Brassica rapa* may grow and defend. *Ecology* 83, 505-517.

Sikora, E., Cieslik, E., Leszczynska, T., Filipiak-Florkiewicz, A., Pisulewski, P.M. (2008). The antioxidant activity of selected cruciferous vegetables subjected to aquathermal processing. *Food Chem.* 107(1), 55–59.

Sikora E, Cieslik E, Filipiak-Florkiewicz A, Leszczyńska, T., (2012). Effect of hydrothermal processing on phenolic acids and flavonols contents in selected brassica vegetables. *Acta Sci. Pol., Technol. Aliment.* 11(1) 2012, 45-51.

Singh, J., Upadhyay, A.K., Prasad, K., Bahadur, A., Rai, M., (2007). Variability of carotenes, vitamin C, E and phenolics in *Brassica* vegetables. *J. Food. Compos. Anal.* 20(2), 106-112.

Smirnoff, N., (1995). Antioxidant Systems and Plant Response to the Environment. *In*: Environment and Plant Metabolism: Flexibility and Acclimation, Smirnoff, N. (Eds.). Bios Scientific Publishers, Oxford, pp: 243-317.

Smith, A.M., Baghurst, K.I., (1992). Public health implications of dietary differences between social status and occupational groups. *J Epidemiol Commun H* 46(4), 409-416.

Sones, K., Heaney, R.K., Fenwick, G.R., (1984). The glucosinolate content of UK vegetables: cabbage (Brasica oleracea), swede (B. napus) and turnip (B. campestris). *Food Addit. Contam.* 3, 289–296

Song, L., Thornalley, P.J., (2007). Effect of storage, processing and cooking on glucosinolate content of *Brassica* vegetables. *Food Chem. Toxicol.* 45(2), 216-224.

Sorensen, G., Stoddard, A., Peterson, K., Cohen, N., Hunt, M.K., Stein, E., Palombo, R., Lederman, R., (1999). Increasing Fruit and Vegetable Consumption Through Worksites and Families in the Treatwell 5-a-Day Study. *Am J Public Health* 89(1), 54-60.

Sørensen, H., (1990). Glucosinolates: Structure-properties-function. p. 149-172. *In*: F. Shahidi (Ed.). Canola and rapeseed: production, chemistry, nutrition and processing technology. Van Nostrand Reinhold, New York.

Sosińska, E., Obiedziński, M.W., (2011). Effect of processing on the content of glucobrassicin and its degradation products in broccoli and cauliflower. *Food Control* 22(8), 1348-1356

Sousa, C., Lopes, G., Pereira, D.M., Taveira, M., Valentão, P., Seabra, R.M., Pereira, J.A., Baptista, P., Ferreres, F., Andrade, P.B. (2007). Screening of antioxidant compounds during sprouting of *Brassica oleracea* L. var. *costata* DC. *Comb Chem High Throughput Screen.* 10(5), 377-86.

Sousa, C., Pereira, D.M., Pereira, J.A., Bento, A., Rodrigues, M.A., Dopico-García, S., Valentão, P., Lopes, G., Ferreres, F., Seabra, R.M., Andrade, P.B., (2008). Multivariate analysis of tronchuda cabbage (*Brassica oleracea* L. var. *costata* DC) phenolics: influence of fertilizers. *J. Agric. Food Chem.* 56, 2231–2239.

Sousa, C., Valentão, P., Rangel, J., Lopes, G., Pereira, J.A., Ferreres, F., Seabra, R.M., Andrade, P.B., (2005). Influence of two fertilization regimens on the amounts of organic acids and phenolic compounds of tronchuda cabbage (*Brassica oleracea* L. Var. *costata* DC). *J Agric Food Chem.* 53(23), 9128-9132.

Stables, G.J., Young, E.M., Howerton, M.W., Yaroch, A.L., Kuester, S., Solera, M.K., Cobb, K., Nebeling, L., (2005). Small school-based effectiveness trials increase vegetable and fruit consumption among youth. *J Am Diet Assoc.* 105(2), 252–256.

Stahl, W., Sies, H., (2003). Antioxidant activity of carotenoids. *Mol Aspects Med* 24, 345–351.

Stampfer, M.J., Rimm, E.B., (1995). Epidemiologic evidence for vitamin E in prevention of cardiovascular disease. *Am J Clin Nutr* 62(6), 1365S-1369S.

Stratmann, J. (2003). Ultraviolet-B radiation co-opts defense signaling pathways. *Trends Plant Sci.* 8(11), 526-533.

Talalay, P., Fahey, J.W., Healy, Z.R., Wehage, S.L., Benedict, A.L., Min, C., Dinkova-Kostova, A.T., (2007). Sulforaphane mobilizes cellular defenses that protect skin against damage by UV radiation. *Proc. Natl. Acad. Sci. USA* 104, 17500-17505.

Tang, L., Zhang, Y., Jobson, H.E., Li, J., Stephenson, K.K.,Wade, K.L., Fahey, J.W., (2006). Potent activation of mitochondria-mediated apoptosis and arrest in S and M phases of cancer cells by a broccoli sprout extract. *Mol. Cancer Ther.* 5, 935-944.

Thompson, R.L., Margets, B.M., Speller, V.M., McVey, D., (1999). The health education authority's health and lifestyle survey 1993: who are the low fruit and vegetable consumers? *J Epidemiol Community H* 53, 294-299.

Toivonen, P.M.A., Forney, C., (2004). Broccoli. *In:* The Commercial Storage of Fruits, Vegetables and Florist and Nursery Stock. USDA, ARS Agriculture Handbook #66.

Toma, S., Losardo, P.L., Vincent, M., Palumbo, R., (1995). Effectiveness of beta-carotene in cancer chemoprevention. *Eur J Cancer Prev.* 4(3), 213-224.

Tomás-Callejas, A., Otón, M., Artés, F., Artés-Hernández, F., (2012). Combined effect of UV-C pretreatment and high oxygen packaging for keeping the quality of fresh-cut Tatsoi baby leaves. *Innov Food Sci Emerg Technol* 14, 115–121.

Toner, C., Pitman, S., (2004). Functional Foods: Educating Consumers About Foods with Health Promoting Benefits in Today's Challenging Communications Environment. *Top Clin Nutr* 19(1), 71-78.

Traka, M., Mithen, R., (2009). Glucosinolates, isothiocyanates and human health. *Phytochem. Rev.* 8, 269–282.

Turkmen, N., F. Sari and Y.S. Velioglu, (2005). The effect of cooking methods on total phenolics and antioxidant activity of selected green vegetables. *Food Chem.* 93, 713-718.

Uchide, N., Toyoda, H., (2011). Antioxidant therapy as a potential approach to severe influenza-associated complications. *Molecules* 16, 2032-2052.

Vallejo, F., Garcia-Viguera C., Tomás-Barberán, F.A., (2003). Changes in broccoli (*Brassica oleracea* L. var. *italica*) health-promoting compounds with inflorescence development. *J. Agric. Food Chem.* 51, 3776 – 3782.

Vallejo, F., Tomás-Barberán, F.A, Ferreres, F., (2004). Characterization of flavonols in broccoli (*Brassica oleracea* L.var. *italica*) by liquid chromatography –UV diode array detection- electro spray ionisation mass spectrometry. *J. Chromatogr. A* 1054, 181-193.

Vallejo, F., Tomas-Barberan, F.A., Banavent-Garcia, A.G., Garcia-Viguera C., (2003).Total and individual glucosinolate contents in inflorescences of eight broccoli cultivars grown under various climatic and fertilization conditions. *J Sci Food Agric* 83, 307-313.

Vallejo, F., Tomás-Barberán, F.A., Garcia-Viguera, C. (2003a). Effect of climatic and sulphur fertilisation conditions, on phenolic compounds and vitamin C, in the inflorescences of eight broccoli cultivars. *Eur. Food Res. Technol.* 216, 395–401.

Vallejo, F., Tomás-Barberán, F.A., Garcia-Viguera, C. (2003b). Phenolic compound contents in edible parts of broccoli inflorescences after domestic cooking. *J Sci Food Agric.* 83(14), 1511–1516.

Vallejo, F., Tomás-Barberán, F.A., Garcia-Viguera, C., (2002). Glucosinolates and vitamin C content in edible parts of broccoli florets after domestic cooking. *Eur Food Res Technol* 215, 310–316.

van Dam, N.M., Raaijmakers, C.E., (2006). Local and systemic induced responses to cabbage root fly larvae (Delia radicum) in *Brassica nigra* and *B. oleracea*. *Chemoecology* 16, 17–24.

van Dam, N.M., Tytgat, T.O.G., Kirkegaard, J.A., (2009). Root and shoot glucosinolates: a comparison of their diversity, function and interactions in natural and managed ecosystems. *Phytochem Rev* 8,171–186.

van Etten, C.H.,. Daxenbichler, M.E, Williams, P.H., Kwolek, W.F., (1976). Glucosinolates and derived products in cruciferous vegetables. Analysis of the edible parts in twenty-two varieties of cabbage. *J. Agric. Food Chem.* 24, 452-455.

Van Poppel, G., Verhoeven, D.T., Verhagen, H., Goldbohm, R.A., (1999). *Brassica* vegetables and cancer prevention. Epidemiology and mechanisms. *Adv. Exp. Med. Biol.* 472, 159–168.

Vanderslice, J.T., Higgs, D.J., Hayes, J.M., Block, G., (1990). Ascorbic acid and dehydroascorbic acid content of foods-as eaten. *J. Food Compos. Anal.* 3, 105–118.

Vaughan, J.G., Geissler, C.A., (1997). The New Oxford Book of Food Plants. Oxford University Press, New York, U.S.A., pp. 166-167.

Velasco, P., Francisco, M., Moreno, D.A., Ferreres, F., García-Viguera, C., Cartea, M.E., (2011). Phytochemical fingerprinting of vegetable *Brassica oleracea* and *Brassica napus* by simultaneous identification of glucosinolates and phenolics. *Phytochem Anal.* 22(2),144-52.

Velasco, P., Soengas, P., Vilar, M., Cartea, M.E., (2008). Comparison of Glucosinolate Profiles in Leaf and Seed Tissues of Different *Brassica napus* Crops. *J. Amer. Soc. Hort. Sci.* 133(4), 551–558.

Verhoeven, D.T., Verhagen, H., Goldbohm, R.A., van den Brandt, P.A., van Poppel, G., (1997). A review of mechanisms underlying anticarcinogenecity by brassica vegetables. *Chem. Biol. Interact.* 103, 79–129.

Verkerk, R. (2002). Evaluation of glucosinolate levels throughout the production chain of *Brassica* vegetables; Towards a novel predictive modelling approach. PhD dissertation,

Wageningen University, Wageningen, The NetherlandsVerkerk, Dekker and Jongen, 2001

Verkerk, R., Dekker, M., (2004). Glucosinolates and myrosinase activity in red cabbage (*Brassica oleracea* L. var. *Capitata* f. *rubra* DC.) after various microwave treatments. *J Agric Food Chem.* 52, 7318–7323.

Verkerk, R., Dekker, M., Jongen, W.M.F., (2001). Post-harvest increase of indolyl glucosinolates in response to chopping and storage of *Brassica* vegetables. *J Sci Food Agric* 81, 953–958.

Verkerk, R., Schreiner, M., Krumbein, A., Ciska, E., Holst, B., Rowland, I., De Schrijver, R., Hansen, M., Gerhäuser, C., Mithen, R., Dekker, M., (2009). Glucosinolates in *Brassica* vegetables: the influence of the food supply chain on intake, bioavailability and human health. *Mol Nutr Food Res.* 53(Suppl 2), S219.

Verkerk, R., van der Gaag, M.S., Dekker, M., Jongen, W.M., (1997). Effects of processing conditions on glucosinolates in cruciferous vegetables. *Cancer Lett.* 114(1-2):193-4.

Vinson, J.A., Dabbagh, Y.A., Serry, M.M., Jang, J., (1995). Plant flavonoids, especially tea flavonols, are powerful antioxidants using an in vitro oxidation model for heart disease. *J. Agric. Food Chem.* 43, 2800–2802.

Volden, J., Bengtsson, G.B., Wicklund, T. (2009). Glucosinolates, L-ascorbic acid, total phenols, anthocyanins, antioxidant capacities and colour in cauliflower (*Brassica oleracea* L., ssp. *botrytis*); effects of long-term freezer storage. *Food Chem.* 112, 967-976.

Volden, J., Wicklund, T., Verkerk, R., Dekker, M., (2008): The kinetics of changes in glucosinolate concentrations during long term cooking of white cabbage (*Brassica oleracea* L. ssp. *capitata* f. *alba*). *J. Agric. Food Chem.* 56, 2068–2073.

Wang, C.Y., (2003). Leafy, floral, and succulent vegetables. *In*: Bartz, J.A. and Brecht, J.K. (Eds.), Postharvest physiology and pathology of vegetables. Second ed. Marcel Dekker, New York, pp. 599-623.

Wei, J., Miao, H.Y., Wang, Q.M., (2011). Effect of glucose on glucosinolates, antioxidants and metabolic enzymes in *Brassica* sprouts. *Sci. Hort.* 129, 535–540.

Wennberg, M., Ekvall, J., Olsson, K., Nyman, M., (2006). Changes in carbohydrate and glucosinolate composition in white cabbage (*Brassica oleracea* var. *capitata*) during blanching and treatment with acetic acid. *Food Chem.* 95, 226–236.

Whitecross, M.I., Armstrong, D.J., (1972). Environmental effects on eEpicuticular waxes of *Brassica-Napus*. *Aust. J. Bot.* 20, 87-95.

WHO/FAO: Diet, nutrition and the prevention of chronic diseases. Tech. Rep. no. 916, Geneva, Switzerland 2003.

WHO: The World Health Report 2002- Reducing Risks, Promoting Healthy Life.

Wills, R.B.H., Rangga, A., (1996). Determination of carotenoids in Chinese vegetables. *Food Chem.* 56, 451-455.

Winkler, S., Faragher, J., Franz, P., Imsic, M., Jones, R., (2007). Glucoraphanin and flavonoid levels remain stable during simulated transport and marketing of broccoli (*Brassica oleracea* var. *italica*) heads, *Postharvest Biol Technol* 43(1), 89-94.

Wu, X., Prior, R.L., (2005). Identification and characterization of anthocyanins by high-performance liquid chromatography-electrospray ionization-tandem mass spectrometry in common foods in the United States: Vegetables, nuts, and grains. *J. Agric. Food Chem.*. 53, 3101–3113.

Wu, L., Ashraf, H.N.A., Facci, M., Wang, R., Paterson, P.G., Ferrie, A., Juurlink, B.H.J., (2004). Dietary approach to attenuate oxidative stress, hypertension, and inflammation in the cardiovascular system. *Proc. Natl. Acad. Sci. USA* 101, 7094-7099.

Yeh, M.C., Ickes, S.B., Lowenstein, L.M., Shuval, K., Ammerman, A.S., Farris, R., Katz, D.L., (2008). Understanding barriers and facilitators of fruit and vegetable consumption among a diverse multi-ethnic population in the USA. *Health Promot Int* 23, 42-51.

Yuan, G.-F., Sun, B., Yuan, J., Wang, Q.-M., (2009). Effects of different cooking methods on health-promoting compounds of broccoli. *J Zhejiang Univ Sci B*. 10(8), 580–588.

Zhan, L.-J.,Wei, G.-Q., Qiao, M.-W., Wang, H.-R., (2012). Quality Improvement of Fresh-Cut Broccoli (*Brassica Oleracea* L.) during Storage by Light Illumination[J]. *Food Science* 33(14), 296-300.

Zhang, D.L., Hamauzu, Y., (2004). Phenolics, ascorbic acid, carotenoids and antioxidant activity of broccoli and their changes during conventional and microwave cooking. *Food Chem.* 88(4), 503–509.

Zhang, H., Schonhof, I., Krumbein, A., Gutezeit, B., Li, L., Stützel, H., Schreiner, M., (2008). Water supply and growing season influence glucosinolate concentration and composition in turnip root (*Brassica rapa* ssp. *rapifera* L.). *J Plant Nutr Soil Sci* 171(2), 255–265.

Zhao, F., Evans, E.J., Bilsborrow, P.E., Syers, J.K., (1994). Influence of nitrogen and sulphur on the glucosinolate profile of rapeseed (*Brassica napus* L). *J. Sci. Food Agric.* 64, 295 – 304.

Zukalová, H., Vašák, J., (2002). The role and effects of glucosinolates of *Brassica* species – a review. *Rostlinná Výroba* 48(4),175-180.

In: Brassicaceae
Editor: Minglin Lang

ISBN: 978-1-62808-856-4

Chapter 3

NEW BROCCOLI VARIETIES WITH IMPROVED HEALTH BENEFITS AND SUITABILITY FOR THE FRESH–CUT AND FIFTH RANGE INDUSTRIES: AN OPPORTUNITY TO INCREASE ITS CONSUMPTION

Ginés Benito Martínez–Hernández[1], Perla A. Gómez[2], Francisco Artés[1,2] and Francisco Artés–Hernández*[*,1,2]

[1]Postharvest and Refrigeration Group, Department of Food Engineering
[2] Institute of Plant Biotechnology,
Universidad Politécnica de Cartagena, Cartagena, Murcia, Spain

ABSTRACT

The health promoting compounds of conventional broccoli varieties have been widely studied in the last years, recommending this vegetable as an excellent source of glucosinolates, phenolic compounds, vitamins, folates, minerals, etc. Thus, its consumption implies many health benefits, including antioxidant and anti–inflammatory properties, regulating enzymes, controlling apoptosis and the cell cycle, coronary diseases prevention, etc. New natural broccoli hybrids, with other *Brassica* species have recently appeared in the market (Bimi®, Bellaverde®, Beneforte®, Broccolini®, etc.) in order to supply milder flavour with even higher health benefits to the consumer, who is still reluctant to the typical sensory quality of conventional varieties. The lifestyle has developed nowadays into a consumer who demands healthy and functional foods with reduced preparation time. In this way, the physiological and physicochemical properties of these new broccoli varieties make them ideal for the fresh–cut and fifth range industries, which can supply new value–added products, ready and easy to eat or cook with high health–promoting value. However, since vegetables processing (fresh–cut, cooking, freezing, etc.) implies changes on their nutritional and bioactive compounds, innovative techniques are appearing with promising results. This chapter reviews the bioactive compounds of these new *Brassicas*, focussing on Bimi®, a new hybrid between broccoli (*Brassica oleracea*, Italica group) and kalian (*B. oleracea,* Alboglabra group), as

[*] E–mail: fr.artes–hdez@upct.es. Web site: www.upct.es/gpostref. Tel: +34–968–325509; Fax: +34–968–325433.

an example of vegetable with great health benefits and high suitability for the fresh–cut and fifth range industries.

Keywords: *Brassica oleracea*; Bimi; minimal processing; cooking; phenolics; Glucosinolates; antioxidant; Vitamins

INTRODUCTION: CLASSIFICATION OF *BRASSICA* SPECIES

The family Brassicaceae is a large group, having about 3,000 species grouped in 350 genera, including several types of edible plants. Economically speaking, the genus *Brassica* is the most important genus within the mentioned family, with 37 different species. This *Brassica* genus includes a group of six interrelated species which was studied by U (1935), who established the relationships among the genomes of these species. The classic Triangle of U is formed by the three diploid species *Brassica nigra* (L.) Koch, *B. oleracea* L. and *B. rapa*. These species naturally hybridized in different combinations to give rise to the three amphidiploids species *B. carinata* A. Braun, *B juncea* (L.) *Czern.* and *B. napus* L. Among these *Brassica* species are oilseed, forage, condiment and vegetable crops by using their buds, inflorescences, leaves, roots, seeds and stems (Soengas *et al.*, 2011). *B. oleracea* is the principal specie, which includes broccoli, kale, cabbage, Brussels sprouts, cauliflower and others.

Broccoli with multiple flower heads is the sprouting broccoli. The term sprouting as used in sprouting broccoli refers to the branching habit of this type, being the young edible inflorescences often referred as sprouts. The sprouting broccoli can be classified according to the colour of the sprouts in green, purple or white sprouting broccoli. On the other side, the broccoli varieties that have a large, single and terminal inflorescence are known as heading broccoli and it can be classified according to its colour in dark purple, copper coloured or purplish–brown, sulphur–coloured or yellowish–green, and green heading broccoli (Giles, 1941). Generally, the green heading broccoli is referred as Calabrese broccoli (*B. oleracea* L. Var. Italica Plenck), which name is derived from its origin located in the Calabria region of Italy. The Calabrese broccoli, or simply known as broccoli, has big dark green–blue florets (10–20 cm diameter) with thick, and tender stems. This broccoli is harvested between late summer and beginning of autumn. In Italy, the term broccoli is used for the edible floral shoots on *Brassica* plants, including cabbages and turnips, and was originally applied to sprouting forms, but now it includes the heading types. The white–heading forms are also colloquially referred to as cauliflower. Broccoli is often used to describe certain forms of cauliflower, notably in the UK where the term heading or winter broccoli is traditionally reserved for biennial types. The term broccoli without qualification is also generally applied in North America to the annual green–sprouting form known in UK and Italy as Calabrese (Dixon, 2007).

Among other common *B. oleracea* varieties are broccoflower® (trademark registered by the company Tanimura & Antle in 1989) and Romanesco broccoli which have characteristic light green heads. However, both forms are considered cultivars of cauliflower *(B. oleracea* var. Botrytis) due to their inflorescence meristems rather than flower buds when harvested. Broccoflower® was originally developed in Holland and has been grown in the USA for nearly twenty years (Tanimura & Antle, 2011). Romanesco broccoli, firstly documented in Italy in the 16th century, has a shape similar to a natural fractal; each bud is composed of a series of smaller buds, all arranged in yet another logarithmic spiral.

HEALTH–PROMOTING PROPERTIES OF BROCCOLI

Clinical trials and epidemiology studies have shown that cruciferous vegetables consumption, and broccoli in particular, reduces the risk of several chronic diseases, such as cardiovascular diseases, inflammation, aging–related disorders and certain types of cancer (Kris–Etheron *et al.*, 2002; Hung *et al.*, 2004; Traka *et al.*, 2008; Bjorkman *et al.*, 2011). In estimating the amount required for this effect, epidemiological studies have revealed that ingestion of three or more half–cup serving of cruciferous vegetables, such as broccoli, Brussels sprouts, or cabbage, per week lower the risk for prostate cancer by 40% compared to ingestion of one or fewer serving per week (Cohen *et al.*, 2000). Bioactive compounds are extranutritional constituents that typically occur in small quantities in foods (Kris–Etherton *et al.*, 2002). Nutritional compounds (macronutrients and micronutrients) are those which the human organism need to develop, in a normal form, the physiological and metabolic processes. The aforementioned health–promoting properties of cruciferous vegetables have been associated, at least in part, to the existence of bioactive compounds, such as phenolic compounds, glucosinolates, carotenoids, chlorophylls and antioxidant enzymes, among others, and other nutritional compounds such as vitamins (C, folic acid, E, K, etc.), fatty acids, minerals, etc.

Table 1. Main broccoli glucosinolates (*cvs.* Emperor, Shogun, Marathon and Viola). Elaborated from Schonhof *et al.* (2004)

Trivial name	Systematic name	Class	Content mg 100 g^{-1} fw (% relative abundance)
Glucoraphanin	4–Methylsulphinylbutyl	Aliphatic	6.6–33.9 (27.6–52.3)
Glucobrassicin	3–Indolylmethyl	Indolyl	6.9–12.1 (11.9–42.6)
Gluconapin	3–Butenyl	Alkenyl	1.0 (1.6)
Proigoitrin	(2R) 2–Hydroxy–3–butenyl	Alkenyl	0.6–10.3 (3.7–15.8)
4–Methoxyglucobrassicin	4–Methoxy–3–indolylmethyl	Indolyl	0.4–1.3 (1.4–3.9)
4–Hydroxyglucobrassicin	4–Hydroxy–3–indolylmethyl	Indolyl	0.06–0.36 (0.4–1.0)
Neoglucobrassicin	N–Methoxy–3–indolylmethyl	Indolyl	0.9–6.3 (3.1–9.1)
Sinigrin	2–Propenyl	Alkenyl	0.9 (1.5)
Glucoiberin	3–Methylsulphinylpropyl	Aliphatic	0.3–3.8 (1.6–14.8)

Glucosinolates and Isothiocyanates

The glucosinolates, previously known as mustard oils (discovered in the 17th century in the mustard), are sulphur–containing compounds mainly found in the Brassicaceae Family. Four types of glucosinolates can be found in the cruciferous vegetables: aromatic (derived from phenylalanine), aliphatic, alkenyl (last two derived from methionine) and indoles glucosinolates (derived from tryptophan) (Wallsgrove and Bennett, 1995). The major glucosinolates present in cruciferous vegetables are 3–butenyl and 4–pentenyl glucosinolates, and their hydroxylated forms that are predominantly found in Chinese cabbage and other forms of *B. rapa* –but also occur in some forms of *B. oleracea*– and 3–methylthiopropyl, 3–methylsulfinylpropyl, 2–propenyl and 4–methylsulfinylbutyl that are found in *B. oleracea*, such as cabbages, cauliflowers and broccoli. The Table 1 shows the main glucosinolates found in broccoli.

The myrosinase (thioglucoside glucohydrolase; EC 3.2.3.1) is largely stored in separate cell compartments from the glucosinolates. When plant cells are damaged (i.e., during food preparation, mastication or injuries caused by predators, such as insects), glucosinolates comes into contact with the enzyme. Then, this enzyme catalyzes the glucosinolates conversion to isothiocyanates (ITCs) after several reactions. Other products of the glucosinolate hydrolysis can be thiocyanates and nitriles, both without bioactive properties, depending on the pH or the presence of metal ions (Fahey and Talalay, 1999). The most bioactive ITCs found in broccoli are sulforaphane (SF) (derived from glucoraphanin), allyl isothiocyanate (derived from sinigrin) and indole–3–carbinol (derived from glucobrassicin) (Jones *et al.*, 2006). When raw florets or sprouts are macerated or eaten, between 60 and 80% of the glucosinolate is converted to the SF–nitrile, as opposed to the ITC, due to the combined effects of myrosinase and a non–catalytic protein cofactor [epithiopsecifier protein (ESP)–like]. However, mild cooking can preserve myrosinase activity while denaturing ESP, resulting in almost 100% conversion to SF. Further cooking denatures myrosinase, and intact glucosinolates are ingested. However, these can be converted to SF in the colon by microbial thioglucosidase activity. It is believed that glucosinolates have no significant biological activity (Fahey *et al.*, 1997). Contrary to them, ITCs are biologically active, typically lipophilic, highly reactive, volatile, malodorous and bitter (Fahey *et al.*, 1999). The physicochemical properties of the ITCs depend upon their characteristic side chains. Thus, certain ITCs, such as phenylethyl, 2–propenyl and 4–methylthiobutyl have a hot flavour, while 3–butenyl and 4–pentenyl ITCs are more pungent. In contrast to the majority of ITCs, SF contributes little to flavour. It is also the most hydrophilic of all the dietary ITCs. Most cultivars of broccoli accumulate between 2 and 10 μmol g^{-1} of 4–methylsulfinyl glucosinolate in their florets. Glucosinolate content is highly influenced by genetics, but its levels can also be altered significantly by the environment that the crop was grown under (Jeffery *et al.*, 2003). Higher levels on a dry weight (dw) basis may sometimes be found within broccoli seedlings ('sprouts') a few days after germination, although these rapidly decline as the seedlings age (Juge *et al.*, 2007).

ITCs have antibacterial and antifungal activities in plants, and provide important protection from insect and herbivore attack (Rosa *et al.*, 1997). A range of ITCs, such as SF, inhibit Phase I enzymes, responsible for the activation of carcinogens, and induce Phase II detoxification enzyme systems, thereby increasing the cancer defence mechanisms of the body, as it has been previously reported *in vitro* (Zhang *et al.*, 1992; Talalay *et al.*, 1995; Munday and Munday, 2004). ITCs have also been implicated in the inhibition of cancer cell proliferation and induction of apoptosis (Musk *et al.*, 1995; Huang *et al.*, 1998; Smith *et al.*, 1998), as well as inhibition of *Helicobacter pylori*, the bacteria responsible for stomach ulcers (Fahey *et al.*, 2002). Furthermore, SF derived from broccoli sprouts has recently been linked to prevention of cardiovascular disease in an animal model study using rats (Wu *et al.*, 2004).

Phenolic Compounds

Phenolics are compounds characterized as having an aromatic ring bearing one or more hydroxyl groups, including their functional derivates (Shahidi and Naczk, 2004). In nature, phenolics are usually found conjugated to sugars and organic acids. Phenolic compounds can be divided in two groups: flavonoids and non–flavonoids. Among flavonoid phenolics (which share a basic structure consisting of two benzene rings linked through a heterocyclic pyrone C ring) are flavonols, flavanones, flavan–3–ols, anthocyanins and isoflavones. In contrast, non–flavonoid phenolics include a more heterogeneous group of compounds including from the simplest of the class, such as C_6–C_1 benzoic acids and C_6–C_3 hydroxycinnamates, to more complex compounds, such as C_6–C_2–C_6 stilbenes, C_6–C_3–C_3–C_6lignans and hydrolysable tannins (gallotannins and ellanditanins) (Shahidi and Naczk, 2004). The phenolic compounds contribute to flavour and colour of the plants. Furthermore, these compounds have many bioactive properties such as antimicrobial, antiviral anti–inflammatory, antitumor activity, anticancer, antimutagenicity, antioxidant potential and reduction in coronary heart disease risk (Lule and Xia, 2005). The most widespread and diverse group of polyphenols in broccoli are the hydroxycinnamic acids, flavonols and anthocyanins. Recently, Redovniković *et al.* (2012) studied the phenolic content of 13 different broccoli *cvs.*, reporting data ranges for total phenolic, and sinapic and caffeic acid derivates of 15.5–26.9, 4.2–7.9 and 0.4–3.2 mg g^{-1} dw. Furthermore, broccoli has been reported as one of the main dietary sources of lignans, comprising coumestans the main group in this *Brassica* (De–Kleijn *et al.*, 2001).

Vitamins

Among broccoli vitamins (C, B1, B2, B5, B6, E, K, PP, etc.), vitamin C is the most studied for its high levels in this vegetable. The majority of vertebrates are able to synthesize vitamin C (L–AA), except a few mammalian species including primates, humans and guinea pigs, being this deficiency localised to a lack of the terminal, flavo–enzyme, L–gulono–1,4–lactone oxidase. L–AA (the AA isomer that occurs in nature) is stable when dry, but solutions readily oxidise, especially in the presence of trace amounts of copper and alkali, may leading to the DHA formation. DHA is unstable itself and undergoes irreversible hydrolytic ring cleavage to 2,3–diketogulonic acid (2, 3–DKG) (without antiscorbutic activity) in aqueous solution. The 2, 3–DKG and D–isoascoric acid (AA stereoisomer) have little, if any, antiscorbutic activity. Generally, three types of biological activity can be defined for L–AA: enzyme cofactor, radical scavenger and donor/acceptor in the electron transport either at the plasma membrane or the chloroplasts (Davey *et al.*, 2000). The L–AA dietary needs refer to a minimum intake of 60 mg day^{-1}, unless it was recommended to be increased to 75 mg day^{-1} for women and 90 mg day^{-1} for men (Padayatty and Levine, 2001). Among vegetables, broccoli is considered as an excellent L–AA source. Many broccoli cultivars have shown levels between 432 and 1,463 mg kg^{-1} fw (Vallejo *et al.*, 2002).

Folate is a generic term for a B–group vitamin, the vitamin B9. There is a large family of naturally occurring folates; mostly reduced tetrahydropteroylglutamates, usually in polyglutamyl form and usually one–carbon substituted (Sanderson *et al.*, 2003). Folate vitamers are present in foods mainly as reduced methyland formyl–tetrahydropteroylpolyglutamates (Perry, 1971; Scott and Weir, 1976). Folic acid (pteroylmonoglutamic acid) is the synthetic form used in supplements and food fortification. Although folic acid fortification was mandatory introduced in the staple foods of some countries like US, Canada and Chile, mandatorial fortification has not yet been implemented in many other countries because of concerns that chronic exposure to excessive doses of folic acid may be mischievous to overall human health (Lucock and Yates, 2005; Kim, 2007). Consequently, dietary folates are predicted to have a higher margin of safety and are currently considered as an alternative for synthetic folic acid (de la Garza *et al.*, 2007). In this way, *Brassicas* are considered as an excellent folate source among vegetables. Broccoli has reported folates ranges of 92–172 μg 100 g^{-1} (Hurdle *et al.*, 1968; McKillop *et al.*, 2002; Houlihan *et al.*, 2011). Folates are themselves not biologically active, but its biological importance is due to tetrahydrofolate and other derivatives after its conversion to dihydrofolic acid in the liver (Bailey and Ayling, 2009).

Fatty Acids

Fatty acids are carboxylic acids with a long aliphatic chain. The ω–3 fatty acids can prevent cardiovascular inflammatory diseases and protect, or even enhance, the effect in medical treatments of diseases like Alzheimer, multiple sclerosis and cancer (Gogus and Smith, 2010). It has been reported the importance of a balanced intake of ω–6 and ω–3 fatty acids, with a ratio of 1–4:1, in order to achieve their health beneficial effects in the control of chronic diseases (Simopoulos, 1999). The major fatty acids reported in broccoli are the two unsaturated fatty acids *α*–linolenic acid [(C18:3, $\Delta^{9,12,15}$) (53 % *v/v*) (ω–3)] and linoleic acid [(C18:2, $\Delta^{9,12}$) (18 % *v/v*) (ω–6)], and the saturated palmitic acid (C16:0, 16 % *v/v*). Broccoli also have small amounts of other unsaturated fatty acids (C16:1, C16:2, C16:3 and C18:1) and the saturated stearic acid (C18:0) (Page *et al.*, 2001). The total fatty acid content reported by Page *et al.* (2001) in broccoli (cv. Marathon) was 34.2 μmol g^{-1} fw, amounting *α*–linolenic and linoleic acids (the majority polyunsaturated fatty acids of broccoli) approximately a 19 and 53 %, respectively, of the total fatty acid content. Similarly, Zhuang *et al.* (1997) reported that the total polyunsaturated acid content of broccoli (*cv*. Iron Dukc) represented a 68 % of the total fatty acid content.

Chlorophylls and Carotenoids

Colours in fruits and vegetables are mainly due to three families of pigments, chlorophylls, carotenoids and anthocyanins, responsible of green, red–yellow and red to blue–purple colours, respectively (Artés *et al.*, 2002). Chlorophylls are natural pigments found in the chloroplasts of plants, algae and some cyanobacteria. Several factors such as cultivar, preharvest conditions, postharvest handling and processing may vary the chlorophyll content of vegetables. In this way, there is a wide range of the chlorophyll content previously reported in broccoli with 260 to 970 mg kg^{-1} fw, reporting chlorophyll *a* around 3–fold higher levels than chlorophyll *b* (Funamoto *et al.*, 2002; Lemoine *et al.*, 2008). Contrary, kailan reported 4.6–fold higher chlorophyll *b* content than *a* over a total chlorophyll content of approximately 400 mg kg^{-1} fw (Noichinda *et al.*, 2007). Chlorophylls have been associated with many health benefits, such as anti–inflammatory, antioxidant, anticancer properties and kidney stones prevention (Tawashi *et al.*, 1982; Dashwood *et al.*, 1998; Egner *et al.*, 2003; Lanfer–Marquez *et al.*, 2005).

Carotenoids are natural pigments that can be classified into two classes: xanthophylls and carotenes. Xanthophylls contain oxygen in their structure while carotenes are purely hydrocarbons, which have not oxygen. The xanthophyll group includes, among others, lutein, zeaxanthin, neoxanthin, violaxanthin, and α– and β–cryptoxanthin. The α-, β- and γ-carotenes are the main carotene types, which are very important as precursors of the vitamin A (Maiani *et al.*, 2009). Only β–carotene, β–cryptoxanthin, α–carotene, lycopene, lutein and zeaxanthin represent more than 95 % of total blood carotenoids in humans. Lutein, zeaxanthin and other xanthophylls are believed to function as protective antioxidants in the macular region of the human retina. The latter carotenoids have also shown protection against cataract formation, coronary heart diseases and stroke (Snodderly, 1995; Ribaya–Mercado and Blumberg, 2004; Chrong *et al.*, 2007). β–carotene has shown other health benefits, may be related to their antioxidative potential, such as enhancement of the immune system function (Bendich, 1989), protection from sunburn (Mathews–Roth, 1990) and inhibition of the development of certain types of cancers (Nishino, 1998). The main carotenoids found in broccoli are lutein and β–carotene with ranges of 707–3,300 and 291–1,750 μg 100 g^{-1} fw, respectively (Leth *et al.*, 2000; Murkovic *et al.*, 2000; Larsen and Christensen, 2005).

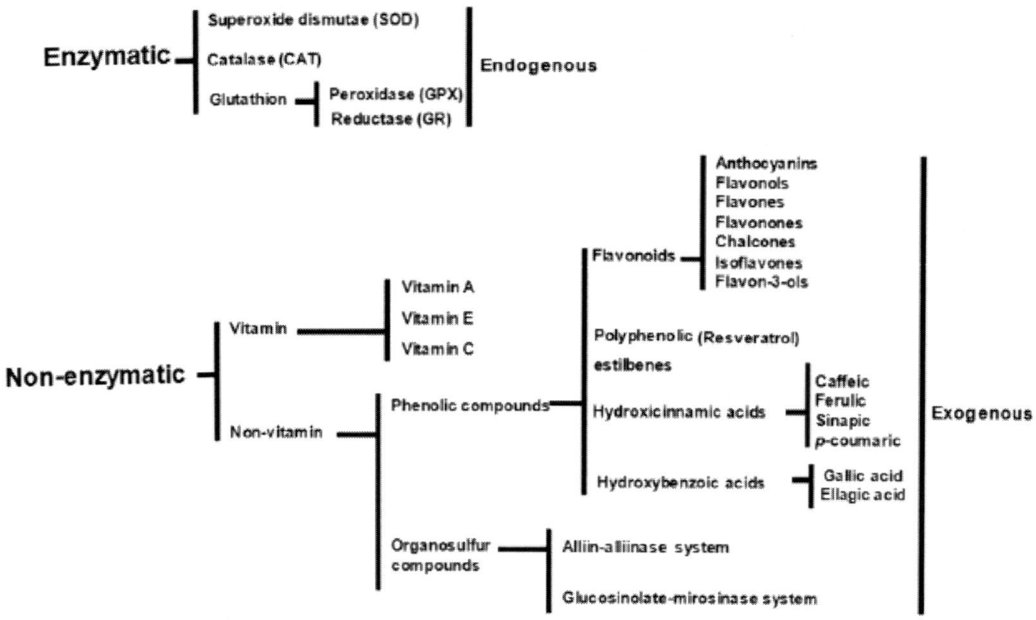

Figure 1. Antioxidant compounds classification.

Oxidative Stress and Antioxidant Capacity

The reactive oxygen species (ROS) are chemically reactive molecules containing oxygen. The ROS includes oxygen ions (i. e., $^1O_2$), free radicals (O_2^-, $\cdot OH$, $NO\cdot$, etc.) and peroxides (H_2O, $ONOO^-$, etc.). ROS are highly reactive due to the presence of unpaired valence shell electrons. ROS form as a natural by–product of the normal metabolism of oxygen and have important roles in cell signalling and homeostasis. However, under exogenous (heat exposure, ultraviolet light, ozone, contaminants, additives, tobacco, drugs, etc.) or endogenous stresses (monoelectronic O_2 reduction, autoxidation of carbon compounds, catalytic activation of several enzymes, etc.), ROS levels can increase dramatically. This may result in significant damage to cell structures. Cumulatively, this is known as oxidative stress. In this way, the antioxidants are compounds that at low concentrations, compared to the substrate, significatively delay or prevent the oxidation of that substrate during an oxidative stress (Devasagayam *et al.*, 2004). According to its nature, these compounds may be classified as enzymatic or non–enzymatic antioxidant compounds (Figure 1). The TAC can be influenced by physiological (i.e., ripening, senescence) and technological factors (storage and processing conditions).

Minerals

Minerals are micronutrients that, although they yield no energy, are necessary for the maintenance of certain essential physicochemical processes of the human metabolism (Soetan *et al.*, 2010). Minerals may be broadly classified as macro (major) or micro (trace) elements. The macroelements (Na, K, Ca, Mg, Cl, P and S) are essential for human being in amounts above 50 mg day^{-1} while microelements (Fe, Zn, Cu, Mn, I, F, Se, Cr, Mo, Co and Ni) are essential in concentrations of below 50 mg day^{-1} (Moreno *et al.*, 2006). Broccoli is a good source of all the macro elements and some important microelements, such as Fe, Zn, Mn and Cu (Table 2). The mineral composition varies among cultivars and is also influenced by several preharvest factors (climate conditions, cultural practices, etc.) (Rosa *et al.*, 2002).

Table 2. Mineral content ranges among different broccoli cultivars (Extracted from Rosa *et al.*, 2002; López–Berenguer *et al.*, 2007)

Mineral	Content
Na	2.0–3.2^{a}
K	23.9–33.8^{a}
Ca	2.1–6.8^{a}
Mg	1.3–2.0^{a}
Cl	4.1–54.0^{a}
P	4.9–9.6^{a}
S	8.9–18.7^{a}
Fe	79.35–96.1^{b}
Zn	35.2–52.8^{b}
Mn	30.1–32.4^{b}
Cu	4.0–13.0^{b}

^{a} mg g^{-1} dw
^{b} µg g^{-1} dw

Dietary Fiber and Proteins

The dietary fiber is mainly composed by macromolecules such as the cellulose, hemicellulose, pectins, lignin, resistant starch and oligosaccharides non–digestible. Broccoli is a rich dietary source of dietary fiber with values around 3.0 % fw (Souci *et al.*, 2000). However, the dietary fiber content can greatly vary among different broccoli varieties.

In animals, amino acids are obtained through the consumption of foods containing proteins. Generally, the protein content of fruit and vegetables is less than 1 % (fw). However, broccoli is considered as an excellent protein source with quantities between 1 and 4%, depending of the variety (Zhuang *et al.*, 1994; Souci *et al.*, 2000).

New Broccoli Varieties

In the last years, broccoli has become in a vegetable with high scientific interest due to the mentioned nutritional and health–promoting properties. The interest in this vegetable has lead to the breeding companies to conduct an intensive genetic improvement of broccoli in order to satisfy the market needs and, in the same way, to the different production areas.

New Varieties with Improved Preharvest Properties

Broccoli generally grows best in cool conditions (i.e., less than 23 °C), with moderate day temperatures and cool nights, to induce and maintain vernalization and to allow normal floral and head development to proceed. High temperatures arrest the inflorescence development in broccoli. Furthermore, broccoli is very sensitive to relatively short periods of heat stress thereby making field observations too variable for effective genetic screening (Björkman and Pearson, 1998). Consequently, much of the production occurs in spring and fall. Several current breeding programs, such as the Eastern Broccoli Project (Cornell University, USA), promise more versatility with new breeding lines for adaptation to summer conditions in the Southeast of USA (hot days and hot nights).

Other broccoli varieties have been developed in order to favour disease (particularly to downy mildew) and insect resistance. Early attempts to introduce black rot (the major seed–borne broccoli disease, caused by *Xanthomonas campestris* pv *campestris*) resistance from *B. carinata* into *B. oleracea* were made using somatic hybridization, and recently *in vitro* embryo culture was used to introgress resistance (Tonguç and Griffiths, 2004). Although a number of diseases may affect broccoli regionally, head rot, caused by a complex of soft rot bacteria (*Erwinia* and *Pseudomonas* spp.), can cause problems whenever water accumulates on the developing broccoli head. Genetic variation in head rot resistance exists in broccoli and is associated with smooth, domed heads and small, tight beads (Darling *et al.*, 2000). Blackleg (*Leptosphaeria maculans*, formerly *Phoma lingam*), and Alternaria (caused by various *Alternaria* spp., but mainly *A. brassicola*) are two diseases that cause significant economic losses in Europe and eastern USA where pesticide–based control options used by conventional growers are not available to organic growers (Lammerts van Bueren *et al.*, 2002). Differences in genetic resistance have been observed among various *Brassica* species, and this resistance needs to be transferred into a *B. oleracea* background (Lammerts van Bueren *et al.*, 2011).

Open pollinated broccoli varieties for organic production have been developed by the Oregon State University using a farmer participatory approach. Current efforts are focused on working with specific farmers and institutions to reduce the variability in the population for economically important traits using plant to row half–sib selection with the intention of developing varieties that are specifically adapted to grower's site–specific conditions (Lammerts van Bueren *et al.*, 2011).

However, the open pollinated cultivars may degenerate due to continuous selfing. In this way, pollination control mechanism, such as the cell fusion for introducing cytoplasmic male sterility (CMS) from other species, have been applied to ease the F1 hybrid production in *Brassicas* species (Lammerts van Bueren and Haring, 2009). Somatic hybridization has been used to transfer the *B. oleracea* nucleus into radish cytoplasm (Budar and Pelletier, 2001), in order to achieve the most widely used form (Ogura) of CMS. The original Ogura CMS was not economically useful because the CMS lines exhibited low temperature chlorosis. It was not until further *in vitro* manipulation that replaced the radish chloroplast genome with the original parental species that temperature insensitive CMS lines were developed (Hoekstra *et al.*, 2010). Many broccoli hybrids currently on the market are produced on male sterile mother plants derived from such cell fusion (Billmann, 2008), and it is difficult for growers to obtain information on the breeding history as declaration is not mandatory. Broccoli is relatively easy to culture *in vitro* and can be transformed. It is also possible to produce doubled haploids through another culture (Lammerts van Bueren *et al.*, 2011).

New Varieties with Improved Health Benefits

Variation in levels of glucosinolates, vitamins, carotenoids, flavonoids, other antioxidant compounds, etc. between and within each group of *Brassica* vegetables suggests differences in their health–promoting properties. Furthermore, the diversity in these nutritional and bioactive compounds is greatly dependent on the cultivar selected. Among preharvest factors, the genetic variation of the cultivar may be regarded as the main factor involved in the varying contents of the phytochemicals of broccoli and other *Brassicas* (Van Etten *et al.*, 1976; Fenwick, 1987; Rosa, *et al.*, 1996; Schonhof, *et al.*, 1999; Rosa and Rodrigues, 2001; Farnham *et al.*, 2005). This diversity indicates that potential health benefits depend greatly on the genotype consumed. The variability of each compound between or within a subspecies is relevant because it can be used to estimate the potential maximal concentration of each compound that can be achieved through genetic manipulation. The greater the variability for a specific trait, the greater the opportunity for genetic improvement by plant breeding (Kurilich *et al.*, 1999). In addition to genetic regulation, there are many indications that environmental and agronomic factors such as water availability (irrigation), soil composition (mineral and organic nutrients), fertilization (sulphur, nitrogen, etc.), light intensity and quality (wavelength), and seasons (daylight and photoperiods), among others (Rossiter and Barrow, 1972; McClure, 1975; Dussi *et al.*, 1995; Falconer and Mackay, 1996; Sarikamis *et al.* 2006).

According to this, Vallejo *et al.* (2003) studied the effect of extreme agronomic and environmental conditions (late season and rich sulphur fertilisation which could induce different stress situations on the plant) on five commercial (Vencedor, Furia, Pentathlon,

Monterrey and Marathon) and three experimental broccoli lines. The phenolic content was enhanced after these growing conditions. Thus, total flavonoids showed the highest content, followed by total sinapic and feruloyl acid derivatives and total caffeoyl–quinic acid derivatives. In general, cultivars grown under rich fertilisation and late season conditions showed higher vitamin C content than those grown under the poor and early ones. Attending to differences among broccoli varieties, these results showed that commercial cultivars rendered higher amounts of phenolic compounds and vitamin C than the experimental ones.

Recently, Farnham and Kopsell (2009) studied the relative effects of genotype versus environment in influencing carotenoids and chlorophylls levels of nine inbred broccoli lines. Concretely, these authors found the lutein as the major carotenoid with a range of 65.3–139.6 $\mu g\ g^{-1}$ dw, showing genotype a highly significant effect (ratio $\sigma^2_g/\ \sigma^2_p = 0.84$). Violaxanthin also exhibited a significant genotype effect, but it was found at lower levels (17.9 to 35.4 $\mu g\ g^{-1}$ dw) than lutein. *β*–carotene and neoxanthine were detected at levels similar to violaxanthin, but genotypic differences were not detected when all environments were compared. This was also true for antheraxanthin, which was detectable in all genotypes at lower levels (mean of 13.3 $\mu g\ g^{-1}$ dw) than the other carotenoids. Attending to chlorophylls, significant genotypic differences were observed for both chlorophylls a and b among the studied inbreds; however, no environment or genotype–by–environment effects were observed with these compounds. As it was observed most carotenoids were positively and significantly correlated with one another, indicating that higher levels of one carotenoid were typically associated with higher levels of others. This study emphasized the relative importance of lutein in broccoli heads and the key role that genotype plays with this compound, ultimately indicating that breeding cultivars with increased levels of this particular carotenoid may be feasible. An examination of 50 broccoli varieties showed that *β*–carotene levels varied over a six–fold, although the *α*–tocopherol and ascorbate variations were not in concert with the *β*–carotene (Kurilich *et al.*, 1999).

The genetic improvement of food crops is no longer constrained by traditional hybridization techniques. Genetic modification via transformation can produce varieties that fall outside the nutrient range found among traditional varieties. Some breeding programmes already select for improved contents of these nutritionally desirable compounds. Previous studies have suggested that enhancing the level of glucosinolates in cruciferous vegetables through conventional breeding or genetic engineering can be expected to enhance the chemopreventive properties of these vegetables. For example, glucosinolate levels in commercially grown broccoli are relatively low compared with those found in salad crops such as rocket (*Eruca sativa*), which accumulates 4–methylthiobutyl glucosinolate, and watercress (*Rorippa nasturtiumaquaticum*), which accumulates phenylethyl glucosinolate (Fenwick *et al.*, 1983a). In this way, new broccoli varieties with high glucosinolate content are being developed as it is described in the following paragraphs.

Redovniković *et al.* (2012) showed a significant variation in the levels of bioactive compounds, and consequently potential health benefits, of 13 broccoli cultivars (Chevalier, Lucky, Belstar, Fiesta, Marathon, Heraklion, Green Magic, Agassi, Montop, Captain, Ironman, General and Parthenon). In this way, the following ranges were observed: total glucosinolates 12.04–22.48 $\mu mol\ g^{-1}$ dw; total phenolic content 15.54–26.92 $mg\ g^{-1}$ dw; and total carotenoid content 0.19–0.46 $mg\ g^{-1}$ dw. Focusing on each cultivar of the latter study, the *cv.* Marathon showed the highest total phenolic and flavonoids, and sinapic and caffeic acid derivates with 26.9, 4.1, 9.1 and 3.2 $mg\ g^{-1}$ dw, respectively. This broccoli cultivar also showed a high total carotenoid content of 0.3 mg g^{-1} dw, unless the *cv*. General registered the highest total carotenoids content with 0.5 mg g^{-1} dw. Furthermore, Redovniković et al. (2012) reported the highest total glucosinolate content for the *cvs*. Ironman, General and Marathon with 22.5, 21.8 and 21.6 µmol g^{-1} dw, respectively, without significant differences among them. Among the individual glucosinolates profiles of these 13 broccoli cultivars, the Marathon reported the highest glucoiberin and glucoraphanin content with 0.9 and 4.1 µmol g^{-1} dw, respectively. Meanwhile, the *cv*. Ironman reported the highest glucobrassicin and 4–methoxyglucobrassicin content with 13.2 and 0.8 µmol g^{-1} dw, respectively.

In a previous study, twelve experimental broccoli lines (open–pollinated varieties) were compared to the commercial *cvs*. Marathon and Lord (Vallejo *et al*., 2002). The experimental lines showed higher total glucosinolates and vitamin C levels, up to 5–6 and 2–3–folds, respectively, than the commercial lines. However, the commercial lines showed caffeic and sinapic acid derivates, and flavonoids values approximately 4, 2.5, and 5–folds higher than experimental lines, respectively.

Fifty different broccoli accessions were studied in a previous research (Kushad *et al*., 1999), reporting glucoraphanin levels ranging from 0.8 µmol g^{-1} dw in EV6–1 to 21.7 µmol g^{-1} dw in Brigadier. Concentrations of the other glucosinolates in broccoli varied similarly over a wide range.

Broccoli hybrid lines developed from crosses between wild *Brassicas* species and broccoli accessions reported 10–fold higher glucoraphanin levels (Faulkner *et al*., 1998). Among the wild *Brassica* species, *B. villosa* showed a great total glucosinolate content with 124 µmol g^{-1} dw, reporting glucoiberin the highest level over the glucosinolates found in the species with 119 µmol g^{-1} dw. The hybrid line GD DH x *B. villosa* from this study reported the highest glucoiberin and glucoraphanin levels with 26.4 and 81.8 µmol g^{-1} dw, respectively. Subsequently, Mithen *et al*. (2003) described the use of these hybrids to develop broccoli breeding lines and experimental hybrids with enhanced levels of glucoraphanin and glucoiberin. Furthermore, Sarikamis *et al*. (2006) suggested that while the absolute levels of glucosinolates and their derivatives are influenced by both genetic and environmental factors, this high–glucosinolate broccoli produced about three fold greater levels of glucosinolates than standard broccoli when grown at different sites and in different years. These authors reported that a mild cooking treatment (microwaving for 1.5 min) generated about three fold higher levels of SF in this high–glucosinolate broccoli than conventional varieties. The raw tissue of high–glucosinolate breeding lines showed a glucosinolate to ITC conversion of 89% compared to commercial *cvs*. Marathon (73%) and Iron (69%). This conversion was analysed after a microwaving treatment for 1.5 min, which allowed destroying the ESP–like protein but did not denature the myrosinase. Cooking for longer would denature myrosinase in both *cv*. Marathon and high–glucosinolate broccoli, and generation of ITC *in vivo* would depend upon thioglucosidase activity of the colonic microflora. Furthermore, a commercial freezing and storage treatment (–20 ºC for 8 weeks) of high glucosinolate broccoli maintained the high level of glucosinolates compared to standard cultivars, although the previous blanching process (90.5 ºC for 80 s) denatured the endogenous myrosinase activity (Sarikamis *et al*., 2006). Furthermore, the ITC–enriched broccoli had 80–times the ability to induce quinone reductase (a standard assay of phase II induction potential) when compared to standard commercial broccoli, due both to an increase in the precursor glucosinolates and a greater conversion of these into ITCs (Mithen *et al*., 2003). A subsequent study using the mentioned high–glucosinolate broccoli, conducted by the same research group, provided, for the first time, experimental evidence obtained in humans to support observational studies that diets rich in cruciferous vegetables may reduce the risk of prostate cancer and other chronic disease (Traka *et al.*, 2008). This 'super broccoli' is sold under the commercial name Beneforté® (Seminis Vegetable Seeds, Inc.) as a fresh–cut product. The first year it was sold in UK and lately it was commercialized in other worldwide areas.

Bitter–tasting foods are frequently disliked and are one reason for low acceptability of *Brassica* (Drewnowski *et al.*, 1999). The literature describes a bitter effect mainly for sinigrin, gluconapin and progoitrin, but also for glucobrassicin and neoglucobrassicin, but in different intensities (Fenwick *et al.*, 1983b; Hansen et al., 1997; Van Doorn et al., 1998; Engel *et al.*, 2002; Schonhof *et al.*, 2004). Besides catabolic products of glucosinolates, other aromatic substances also play a role, such as hexanal, (E)–2–hexenal, methanethiol, hydrogen sulfide, dimethyl disulfide, dimethyl trisulfide, and S–methylcysteine sulfoxide, among others (Maruyama, 1970; Forney *et al.*, 1991; Hansen *et al.*, 1992; Marks *et al.*, 1992; Ulrich *et al.*, 1998). Other volatile compounds, such as ethanol, ethyl acetate, acetaldehyde, methyl acetate, and acetone (Faulkner *et al.*, 1998; Drewnowski *et al.*, 1999), have also been found to be responsible for off–odours during anaerobic respiration. Furthermore, sweet and bitter taste may be very closely related (Walters, 1996; Walters and Roy, 1996). The sweet impression seemed to be decreased with increasing contents of bitter glucosinolates. In this way, the broccoli *cv.* Shogun with a very high level of bitter glucosinolates, and at the same time, lower sucrose content, was rejected (Schonhof *et al.*, 2004). According to this, a higher level of sucrose concentration seemed to lead to a masking of the bitter glucosinolates with a positive influence on consumer acceptability.

The consumer acceptability of broccoli has been reported to be mainly influenced by flavour and colour of the product (Schonhof *et al.*, 2004). A successful marketing or a targeted breeding strategy for broccoli has to consider obtaining cultivars which mask the bitter taste of some glucosinolates by raising sugar content (Schonhof *et al.*, 2004; Walters and Roy, 1996). In this way, new broccoli cultivars with sweeter taste, such as the natural hybrid between the kailan and broccoli, have appeared. This new kailan–hybrid broccoli, described in the following paragraphs, is an excellent *Brassica cv.* for the fresh–cut (FC) and fifth range industry.

Kailan–hybrid Broccoli

The kailan–hybrid broccoli (*B. oleracea* Italica x Alboglabra group) is a vegetable with remarkable more pleasant flavour and aroma than the conventional broccoli varieties, and presumably with a bioactive and nutritive profile similar to that of broccoli and kailan. It is characterized by a floret at the end of each stem (Figure 2). Similarly to broccoli, kailan–hybrid broccoli has yellow flowers.

This hybrid was firstly developed by the Sakata Seed Company of Yokohama (Japan). Other companies have developed different commercial kailan–hybrid broccoli varieties with registered trademarks: Bimi® (Sakata Vegetables Europe), Asparation® (Sakata Seed America), Bellaverde® (Seminis Vegetable Seeds), Broccolini® (Mann Packaging Company), Tenderstem® (Marks and Spencer Plc.), among others. This vegetable was firstly commercialized by Sabon Incorporated, which made a commercial program to sell Asparation® in México in 1994. Mann Packing Company introduced the new vegetable to the

USA market in 1998. Lately, its cultivation has been extended to the other countries around the world, such as the northern European countries, Brazil, Australia, etc. This broccoli has become very popular in some countries like Brazil, where it is simply known as 'broccoli', referring to the heading broccoli as American broccoli. In that country, the production of the American broccoli is low compared to that of the kailan–hybrid broccoli, being the common varieties the 'Ramoso de Piracicaba' and 'Ramoso Santana' (Cenci and Gomes, 2007).

Figure 2. Visual morphological differences and different parts of broccoli *cv*. 'Parthenon' (right) and kailan–hybrid broccoli (left).

Due to its delicate physical properties, is manually collected every day (in the first hours of the morning). In order to extend the postharvest shelf–life of the kailan–hybrid broccoli by minimizing the handling, its primary packaging is made on the fields. Kailan–hybrid broccoli is manually harvested due to the sensitivity to damages of this crop and mechanization difficulties. The following manual harvesting forms are commonly used:

- Whole head: all stems are used with good uniformity.
- Advanced stems from the main head: only the most advanced stems are cut and the rest are harvested in the following harvest time.
- Regrowths.

The maturity indices for conventional broccoli are head diameter (different for every variety) and compactness. Furthermore, all florets should be closed. The optimum maturity indices of kailan–hybrid broccoli at harvest, which will ensure a good quality of the FC produce, are:

- No floret yellowness.
- Stem Form: First category (without nodes) and second category (four nodes maximum).
- Diameter: 8–15 mm.
- Long: 120–220 mm.

The mild sensory characteristics of the kailan–hybrid broccoli, compared to the conventional broccoli varieties, allow to eat it either raw (i.e., in salads) or cooked. The kailan–hybrid broccoli can be cooked with mild cooking procedures due to its tenderness and small size (better heat transmission during cooking). This is the great advantage that drastically reduces the inevitable nutritional losses during the cooking treatments, which are usually conducted for the consumption of the conventional broccoli varieties due to their characteristic bitter and astringent flavours.

The production of this vegetable is concentrated from October to June in warm areas, while in the summer months it is produced in northern locations as UK, The Netherlands, etc. However, commercial production is mainly located in Africa, where production is assured all year round with intensive farming systems. Spain is among the main kalian-hybrid producers with a cultivated surface (principally concentrated in the south-east area) of 16 ha and a production of 65 t in the campaign 2010/11 (data supplied by Sakata Seed Ibérica). Kailan–hybrid consumption has already begun in many European countries, such as Belgium, UK, France, Germany, The Netherlands and the Scandinavian countries, but so far it does not exist in the Spanish market.

The actual consumer, with scarce time to prepare a convenient meal (which usually does not meet the nutritional dietary requirements), looks for convenient fresh foods with no additives, high nutritional value and antioxidant properties to be consumed both at home and food services (Artés *et al.*, 2009). In order to supply these consumer requirements, the FC or minimally processed vegetables have appeared in the last years. The FC vegetables are ready–to–eat products, elaborated free from additives by using light combined methods such as peeling, cutting, washing, sanitation, rinsing, dewatering, packaging (usually under modified atmosphere packaging, MAP) and storage under refrigerated conditions (Artés and Allende, 2005). These products usually do not need further processing prior to consumption.

Attending to the physiological properties, the kailan hybrid has higher respiration rate and ethylene production than conventional broccoli cultivars. The MAP, combined with low storage temperature, allows extending the shelf–life of this vegetable as a FC product. In this way, a MAP storage at 2 ºC with 5–7 kPa O_2 + 14–15 kPa CO_2 (balanced with N_2) allowed to reduce the mesophilic, psychrophilic, enterobacteria and yeast and mould counts in a 43, 10, 57 and 10 % after 15 days, respectively, compared to control samples stored in air. After 15 days at 2 and 5 ºC under MAP, undesirable sensory attributes changes, such as yellowness, off–flavour, and stem softening and bent, were retarded. In air–stored samples at 8 ºC a reduction in total soluble solids content and an increase in yellowing were found. MAP storage at 2 ºC (5–7 kPa O_2 + 14–15 kPa CO_2) or 5 ºC (1.5–2.5 kPa O_2 + 15–16 kPa CO_2) provided an acceptable sensory quality and safety after 15 days. Attending to nutritional properties, kailan–hybrid broccoli florets have higher dietary fiber content than stems. Furthermore, the total protein content of kailan–hybrid florets is 2.2–fold higher than conventional broccoli varieties, such that of the 'Parthenon' *cv*. Higher amounts of S, Ca, Mg, Fe, Sr, Mn, Zn and Cu in the kailan–hybrid than those of 'Parthenon' cv. have been reported. However, 'Parthenon' *cv*. florets registered higher initial total phenolics content than the kailan–hybrid broccoli, followed by an increase throughout shelf–life favoured at 5 and 8 ºC under MAP. The MAP of kailan–hybrid broccoli samples and storage at 8 ºC showed higher individual phenolics content than MAP–stored samples at 2 ºC. Our research group have found that the initial total antioxidant capacity (TAC) of the kailan–hybrid was higher than that of 'Parthenon' *cv*. florets. In this way, the kailan–hybrid florets could lead to healthier properties on the mentioned bioactive and nutritive compounds compared to the conventional 'Parthenon' *cv*. (Martínez–Hernández *et al*., 2013a).

Kailan–hybrid broccoli assembles the requirements for an excellent product for the minimal processing or FC. However, the processing and storage of FC products imply important changes on the physicochemical, microbial, sensory, bioactive and nutritional quality of them. Furthermore, the association of foodborne outbreaks with these products has dramatically increased during the last years. In this way, and due to the cancer by–products of the conventionally used NaClO as sanitizer agent, it is imperative to optimise the emerging NaClO–alternative sanitising treatments, such as the electrolysed water (EW), ultraviolet light type C (UV–C) and superatmospheric oxygen packaging (HO), for the kailan–hybrid broccoli. The effects of several UV–C pre–treatments (1.5, 4.5, 9 and 15 kJ m^{-2}) on changes in physiological, sensory and microbial quality, and some bioactive compounds over 19 days at 5 and 10 ºC of FC kailan–hybrid broccoli were studied. Non–radiated samples were used as controls. Low and moderate UV–C doses (1.5 and 4.5 kJ m^{-2}) showed inhibitory effects on natural microflora growth. In relation to sensory quality, all treatments resulted in a shelf–life of 19 and 13 days at 5 and 10ºC, respectively, with the exception of 15 kJ m^{-2} treated samples which resulted in a shorter shelf–life. These doses immediately induced an increase in total polyphenols contents reaching 25 % more after 19 days at 5 ºC compared to the initial value. All the hydroxycinnamoyl acid derivates were immediately increased after UV–C treatments, with values 4.8 and 4.5–fold higher for 4.5 and 9 kJ m^{-2} treated samples, respectively, over the control. Changes in phenolic compounds were highly influenced by the storage temperature throughout shelf–life. The TAC generally followed the same pattern: the higher the UV–C doses, the higher the TAC values. Generally, UV–C slightly reduced initial total chlorophyll content but delayed its degradation throughout shelf–life. In this way, a pre–treatment of 4.5 kJ m^{-2} is useful as a technique to improve epiphytic microbial quality and health–promoting compounds of FC kailan–hybrid broccoli (Martínez–Hernández *et al*., 2011).

The effects of neutral electrolysed water (NEW), UV–C and HO, single or combined, on the quality of FC kailan–hybrid broccoli up to 19 days at 5 ºC were also studied (Martínez–Hernández *et al*., 2013b). After 15 days, the combined treatments achieved lower mesophilic and psychrophilic growth compared to the single ones. Single treatments induced higher ascorbate peroxidase (APX) activity reductions just after its application, while superoxide dismutase activity showed the opposite behaviour. After 5 days at 5 ºC, a great increase of APX and guaiacol peroxidise activity was found, achieving NEW+UV–C+HO and HO–including treatments the highest and the lowest APX activity increases, respectively. UV–C treatments produced the highest α–linolenic acid (ALA) decreases ranging around 35–38 %

over control content on the processing day. NEW treatments greatly reduced, throughout shelf–life, ALA and stearic acid content by 27–44 % and 31–61 %, respectively. Total phenolic content and TAC (1,415 mg chlorogenic acid equivalents–ChAE– kg^{-1}fw and 287 mg ascorbic acid equivalents antioxidant capacity–AAEAC– kg^{-1} fw, respectively) remained quite constant during shelf–life. Consequently, these innovative sanitising treatments, and their possible combinations, seem to be promising techniques for keeping, or even enhancing, the quality of FC kailan–hybrid broccoli and, probably, other new broccoli varieties.

The actual increasing demand of low and easy–preparation foods has attached a considerably long expiration date, usually of several weeks, expected by the consumers. Along the recent and emerging trajectory of the fifth range vegetable industry, conventional cooking techniques, such as boiling, steaming, deep–frying and grilling have been applied. Together with the latter cooking methods, new technologies like vacuum cooking (cook vide), *sous vide* ('under vacuum' from French) and microwaving (MW) are being adapted to the fifth range vegetable industry in order to diversify the product offer, minimise the nutritional losses during thermal treatments, maximize the sensory properties of the product and reduce the production costs (versatile and effective equipment, less energetic cost, etc). These new techniques allow cooking at lower temperature and reduced time in order to obtain products with the aforementioned characteristics. Kailan–hybrid broccoli also has excellent characteristics for the fifth range industry. Nevertheless, the cooking treatments imply important changes on the physicochemical, microbial, sensory, bioactive and nutritional quality of the fifth range products. Among cooking treatments, boiling and steaming produced the greatest stem softening. Based on the overall sensory quality, the commercial life was established in 45 days, except grilling (14 days) and *sous vide* (21 days). Apparently, cooking increased the total phenolic content up to 2.0 and 1.7–fold for grilling and MW, respectively, owing to a better extraction. *Sous vide*–MW, *sous vide* and MW induced the highest TAC increases around 5.4–4.7–fold, contrary to the low enhancements of boiling and grilling (2.9–fold). The best chlorophylls retention was attained by boiling. The total carotenoids content was enhanced up to 1.5–2–fold (Martínez–Hernández *et al.*, 2012a). The glucosinolate profile of kailan–hybrid revealed between 1.7 to 9.3 fold higher glucobrassicin content in their total edible fraction and florets (4.76 and 3.41 mg g^{-1} dw, respectively) than that reported in florets from different conventional broccoli *cvs*. Boiling and *sous vide* induced the highest glucosinolate loss (80%), while low pressure (LP) steaming, MW and *sous vide*–MW showed the lowest (40%) loss. Glucoraphanin was the most thermostable. Throughout their commercial life, microwaved and grilled samples showed a decrease in total glucosinolates. Generally, myrosinase activity was completely inhibited after cooking with undetected sulforaphane contents. The initial total vitamin C dropped by up to 58% after cooking and progressively decreased during storage, with *sous vide*–MW (92%) and MW (21%) showing the highest and lowest decrements, respectively. LP steaming and MW were the best industrial cooking methods for maintaining the glucosinolate and vitamin C contents, and enhancing up to 7.5–fold the initial lutein content (Martínez–Hernández *et al.*, 2013c). *Sous vide*–MW greatly decreased microbial counts, achieving very low psychrophilic and enterobacteria counts (1.1 and 0.2 log CFU g^{-1}, respectively). Vacuum boiling and *sous vide* reduced the stem broccoli firmness by approximately 54–58 %, reaching a pleasant and moderate softening. *Sous vide*, grilling and steaming induced the lowest stem colour changes. Generally, all cooking treatments showed a good overall sensory quality. The total phenolic content (1,148 mg ChAE kg^{-1} fw) usually increased after cooking, with microwave and grilled treatments registering the highest increases up to 2–fold. Commonly, the TAC (296.6 mg AAEAC kg^{-1} fw) increased after cooking by *sous vide*, MW and frying treatments registering the highest increments, by approximately 3.6–fold. Generally, the cooking process reduced the initial vitamin C content, with vacuum and conventional boiling showing the lowest and highest losses with 27 and 62 %, respectively, while vacuum deep frying preserved the initial value (1,737 mg kg^{-1} fw) (Martínez–Hernández *et al.*, 2012b).

Attending to the bioavailability of this broccoli hybrid, preliminary studies in our group point this kailan–hybrid broccoli as a vegetable with great chemopreventive potential after ingestion than other broccoli varieties and *Brassica* species.

CONCLUSION

Broccoli has been described as a 'super–vegetable' among consumers after the numerous epidemiological and laboratory studies on this *Brassica* specie. The health–promoting properties of broccoli shown by these studies are related to its high content in phenolic compounds, glucosinolates (isothiocyanates), carotenoids, vitamins, fatty acids, minerals, etc. In this way, breeding programmes are focusing to produce broccoli varieties with higher nutritional and bioactive properties.

Mild processing methods, such as the fresh-cut and fifth range processing, are preferred by the actual consumer due to the high sensory and nutritional properties of these healthy products with no additives and ready to eat properties. Attending to the cooking step involved in the fifth range processing, microwave, low pressure steaming and vacuum-based treatments achieve the best nutritional compounds retention. In this way, new varieties developed, such as the kalian-hybrid broccoli, which are better adapted to these fresh-cut and fifth range processing, could increase its consumption.

REFERENCES

Artés, F., Allende, A., (2005). Minimal fresh processing of vegetables, fruits and juices. In: Sun, D.W. (Ed.). Emerging Technologies for Food Processing. Elsevier–Academic Press, Oxford, UK. 675–715.

Artés, F., Gómez, P.A., Aguayo, E., Escalona, V.H., Artés–Hernández, F., (2009). Sustainable sanitation techniques for keeping quality and safety of fresh–cut plant commodities. *Postharvest Biol. Technol.* 51: 287–296.

Artés, F., Mínguez, M.I., Hornero, D., (2002). Analyzing changes in fruit pigments. In: Mac Dougall, D.B. (Ed.). Colour in Food. Improving Quality. CRC Press and Woodhead Publishing Ltd., Cambridge, UK. 248–282.

Bailey, S.W., Ayling, J.E., (2009). The extremely slow and variable activity of dihydrofolate reductase in human liver and its implications for high folic acid intake. *Proc. Natl. Acad. Sci. USA.* 106: 15424–15429.

Bendich, A., (1989). Carotenoids and the immune response. *J. Nutr.* 119: 112–115.

Billmann, B., (2008). Impacts of banning protoplast fusion on the range of varieties available for organic arable cropping and vegetable production, FiBL Report as Part of the COOP

Project on Module 1.3. Safeguarding Organic Seed and Planting Material – Impulses for Organic Plant Breeding.

Bjorkman, M., Klingen, I., Birch, A.N.E., Bones, A.M., Bruce, T.J.A., Johansen, T.J., Meadow, R., Molmann, J., Seljasen, R., Smart, L.E., Stewart, D., (2011). Phytochemicals of Brassicaceae in plant protection and human health – Influences of climate, environment and agronomic practice. *Phytochemistry*. 72: 538–556.

Björkman, T., Pearson, K.J., (1998). High temperature arrest of inflorescence development in broccoli (*Brassica oleracea* var.*italica* L.). *J. Exp. Bot.* 49: 101–106.

Budar, F., Pelletier, G., (2001). Male sterility in plants: occurrence, determinism, significance and use. *C. R. Acad. Sci. III, Sci. Vie*. 324: 543–550.

Cenci, S.A., Gomes, C.A.O., (2007). Processamento mínimo de brócolis. In: Moretti, C.L. (Ed.). Manual de Processamento Mínimo de Frutas e Hortaliças. Embrapa Hortaliças e SEBRAE, Brasilia DF, Brazil. 383–392.

Chrong, E.W.I., Wong, T.Y., Kreis, A.J., Simpson, J.A., Guymer, R.H., (2007). Dietary antioxidants and primary prevention of age related macular degeneration: Systemic review and meta–analysis. *BMJ*. 335: 755–759.

Cohen, J.H., Kristal, A.R., Stanford, J.L., (2000). Fruit and vegetable intakes and prostate cancer. *J. Natl. Cancer Inst*. 92: 61–68.

Darling, D., Harling, R., Simpson, R.A., McRoberts, N., Hunter, E.A., (2000). Susceptibility of broccoli cultivars to bacterial head rot: in vitro screening and the role of head morphology in resistance. *Eur. J. Plant Pathol*. 106: 11–17.

Dashwood, R., Negishi, T., Hayatsu, H., Breinholt, V., Hendricks, J., Bailey, G., (1998). Chemopreventive properties of chlorophylls towards aflatoxin B1: A review of the antimutagenicity and anticarcinogenicity data in rainbow trout. *Mutat. Res.* 399: 245–253.

Davey, M.W., Montagu, M.V., Inzé, D., Sanmartin, M., Kanellis, A., Smirnoff, N., Benzie, I.J.J., Strain, J.J., Favell, D., Fletcher, J., (2000). Plant L–ascorbic acid: Chemistry, function, metabolism, bioavailability and effects of processing. *J. Sci. Food Agric.* 80: 825–860.

de la Garza, R., Gregory, J.F., Hanson, A., (2007). Folate biofortification of tomato fruit. *Proc. Natl. Acad. Sci. USA*. 104: 4218–4222.

De–Kleijn, M.J., Van–der–Schouw, Y.T., Wilson, P.W., Adlercreutz, H., Mazur, W., Grobbee, D.E., Jacques, P.F., (2001). Intake of dietary phytoestrogensis low in postmenopausal women in the United States: The Framingham study (1–4). *J. Nutr*. 131: 1826–1832.

Devasagayam, T.P.A., Tilak, J.C., Boloor, K.K., Sane, K.S., Ghaskadbi, S.S., Lele, R.D., 2004. Free radicals and antioxidants in human health: Current status and future prospects. *J. Assoc. Physicians India*. 52: 796–804.

Dixon G.R., (2007). Origins and diversity of *Brassica* and its relatives. In: Vegetable *Brassicas* and related crucifers. Crop production science in horticulture, Vol. 14. CABI publisher, Oxfordshire, UK. 1–35.

Drewnowski, A., Henderson, S.A., Levine, A., Hann, C., (1999). Taste and food preferences as predictors of dietary practices in young women. *Public. Health Nutr*. 2: 503–519.

Dussi, M.C., Sugar, D., Wrolstad, R.E., (1995). Characterizing and quantifying anthocyanins in red pears and effect of light quality on fruit color. *J. Am. Soc. Hortic. Sci*. 120: 785–789.

Egner, P.A., Muñoz, A., Kensler, T.W., (2003). Chemoprevention with chlorophyllin in individuals exposed to dietary aflatoxin. *Mutat. Res*. 523: 209–216.

Engel, E., Baty, C., le Corre, D., Suchon, I., Martin, N., (2002). Flavor–active compounds potentially implicated in cooked cauliflower acceptance. *J. Agric. Food Chem.* 50: 6459–6467.

Fahey, J.W., Haristoy, X., Dolan, P.M., Kensler, T.W., Scholtus, I., (2002). Sulforaphane inhibits extracellular, intracellular, and antibiotic–resistant strains of Helicobacter pylori and prevents benzo[a]pyrene–induced stomach tumors. *P. Natl. Acad. Sci. USA*. 99: 7610–7615.

Fahey, J.W., Talalay P., (1999). Antioxidant functions of sulforaphane: A potent inducer of phase 2 detoxication enzymes. *Food Chem. Toxicol.* 37: 973–979.

Fahey, J.W., Zhang, Y., Talalay, P., (1997). Broccoli sprouts: An exceptionally rich source of inducers of enzymes that protect against chemical carcinogens. *Proc. Natl. Acad. Sci. USA*. 94: 10367–10372.

Falconer, D.S., Mackay, T.F. (1996). Introduction to quantitative genetics. Longman Scientific, Essex, UK.

Farnham, M.W., Kopsell, D.A. (2009). Importance of genotype on carotenoid and chlorophyll levels in broccoli heads. *HortScience*. 44: 1248–1253.

Farnham, M.W., Stephenson, K.K., Fahey, J.W., (2005). Glucoraphanin level in broccoli seed is largely determined by genotype. *HortScience*. 40: 50–53.

Faulkner, K., Mithen, R., Williamson, G., (1998). Selective increase of the potential anticarcinogen 4–methylsulfphinylbutyl glucosinolate in broccoli. *Carcinogenesis*. 19: 605–609.

Fenwick, G.R., Griffiths, N.M., Heaney, R.K., (1983b). Bitterness in Brussels sprouts (*Brassica oleracea* L. var. *gemmifera*): The role of glucosinolates and their breakdown products. *J. Sci. Food Agric*. 34: 73–80.

Fenwick, G.R., Heaney, R.K., Mullin, W.J., (1983a). Glucosinolates and their breakdown products in food and food products. *Crit. Rev. Food Sci. Nutr*. 18: 123–201.

Forney, C.F., Mattheis, J.P., Austin, R.K., (1991). Volatile compounds produced by broccoli under anaerobic conditions. *J. Agric. Food Chem*. 39: 2257–2259.

Funamoto, Y., Yamauchi, N., Shigenaga, T., Shigyo, M., (2002). Effects of heat treatment on chlorophyll degrading enzymes in stored broccoli (*Brassica oleracea* L.). *Postharvest Biol. Technol*. 24: 163–170.

Giles WF., (1941). Cauliflower and broccoli. What they are and where they come from? *Royal Hortic. Soc.* 66: 265–277.

Gogus, U., Smith, C., (2010). n–3 Omega fatty acids: A review of current knowledge. *Int. J. Food Sci. Technol.* 45: 417–436.

Hansen, M., Buttery, R.G., Stern, D.J., Cantwell, M.I., Ling, L.C., (1992). Broccoli storage under low–oxygen atmosphere: identification of higher boiling volatiles. *J. Agric. Food Chem.* 40: 850–852.

Hansen, M., Laustsen, A.M., Olsen, C.E., Poll, L., Sørensen, H., (1997). Chemical and sensory quality of broccoli (*Brassica oleracea* L. var. *Italica*). *J. Food Quality*. 20: 441–459.

Hoekstra, S., Kool, A.J., Nootebos, M.G., Tan, M.–L.M.C., (2010). Male sterile *Brassica* plants U.S. Patent and Trademark Office No. 5254802, 1993; Available from: http://patft.uspto.gov/netahtml/PTO/search–bool.html (verified 23 March 2010).

Houlihan, A., Pyke, M., Scheelings, P., Graham, G., Eaglesham, G., O'Hare, T., Wong, L., Puwastien, P., Jongjitsin, W., (2011). Folate content of Asian vegetables. Rural Industries Research and Development Corporation (Australian Government), Canberra, Australia.

Huang, C., Ma, W.Y., Li, J., Hecht, S.S., Don, Z., (1998). Essential role of p53 in phenethyl isothiocyanate induced apoptosis. *Cancer Res.* 58: 4102–4106.

Hung, H–C., Joshipura, K.J., Jiang, R., Hu, F.B., Hunter, D., Smith–Warner, S.A., Colditz, G.A., Rosner, B., Spiegelman, D., Willett, W.C., (2004). Fruit and vegetable intake and risk of major chronic disease. *J. Natl. Cancer Inst.* 96: 1577–1584.

Hurdle, A.D.F.., Path, M.C., Barton, D., Searles, I.H., (1968). A Method for Measuring Folate in Food and Its Application to a Hospital Diet. *Am. J. Clin. Nutr.* 21: 1202–1207.

Jeffery, E.H., Brown, A.F., Kurilich, A.C., Keck, A.S., Matusheski, N., Klein, B.P., Juvik, J.A., (2003). Variation in content of bioactive components in broccoli. *J. Food Compos. Anal.* 16: 323–330.

Jeffery, E.H., Brown, A.F., Kurilich, A.S., Keck, A.S., Matusheski, N., Klein, B.P., Juvik, J.A., (2003). Variation in content of bioactive components in broccoli. *J. Food Comp. Anal.* 16: 323–330.

Juge, N., Mithen R.F., Traka, M., (2007). Molecular basis for chemoprevention by sulforaphane: A comprehensive review. *Cell. Mol. Life Sci.* 64: 1105–1127.

Kim, Y.I., (2007). Folic acid fortification and supplementation–good for some but not so good for others. *Nutr. Rev.* 65: 504–511.

Kris–Etherton, P.M., Hecker, K.D., Bonanome, A., Coval, S. M., Binkoski, A.E., Hilpert, K.F., Griel, A.E., Etherton, T.D., (2002). Bioactive compounds in foods: Their role in the prevention of cardiovascular disease and cancer. *Am. J. Med.* 113: 71–88.

Kurilich, A.C., Tsau, G.J., Brown, A.F., Howard, L., Klein, B.P., Jeffery, E.H., Kushad, M.M., Wallig, M.A., Juvik, J.A., (1999). Carotene, tocopherol and ascorbate contents in subspecies of *Brassica oleracea*. *J. Agric. Food Chem.* 47: 1576–1581.

Kushad, M.M., Brown, A.F., Kurilich, A.C., Juvik, J.A., Klein, B.P., Wallig, M.A., Jeffery, E.H., (1999). Variation of glucosinolates in vegetable crops of *Brassica oleracea*. *J. Agric. Food Chem.* 47: 1541–1548.

Lammerts van Bueren, E.T., Haring, M., (2009). What is protoplast fusion and what are the objections against protoplast fusion from an organic point of view. In: Rey, F., Gerber, M. (Eds.). Strategies for a future without cell fusion techniques in varieties applied in organic farming. Paris, France, 27–28 April 2009, ITAB and ECO–PB. pp. 5–8.

Lammerts van Bueren, E.T., Jones, S.S., Tamm, L., Murphy, K.M., Myers, J.R., Leifert, C., Messmer, M.M., (2011). The need to breed crop varieties suitable for organic farming, using wheat, tomato and broccoli as examples: A review. *Neth J Agr Sci.* 58: 193–205.

Lammerts van Bueren, E.T., Struik, P.C., Jacobsen, E., (2002). Ecological concepts in organic farming and their consequences for an organic crop ideotype. *Neth J Agr Sci.* 50: 1–26.

Lanfer–Marquez, U.M., Barros, R.M.C., Sinnecker, P., (2005). Antioxidant activity of chlorophylls and their derivatives. *Food Res. Int.* 38: 885–891.

Larsen, E., Christensen, L.P., (2005). Simple saponification method for the quantitative determination of carotenoids in green vegetables. *J. Agric. Food Chem.* 53: 6598–6602.

Lemoine, M.L., Civello, P.M., Chaves, A.R., Martínez, G.A., (2008). Effect of combined treatment with hot air and UV–C on senescence and quality parameters of minimally processed broccoli (*Brassica oleracea* L. var. Italica). *Postharvest Biol. Technol.* 48: 15–21.

Leth, T., Jakobsen, J., Andersen, N.L., (2000). The intake of carotenoids in Denmark, *Eur. J. Lipid Sci. Technol.* 102: 128–132.

Lewis, J., Fenwick, G.R., (1987). Glucosinolate content of *Brassica* vegetables: Analysis of twenty–four cultivars of calabrese (green sprouting broccoli, *Brassica oleracea* L. var. *botrytis* subvar. *cymosa* Lam.). *Food Chem.* 25: 259–268.

López–Berenguer, C., Carvajal, M., Moreno, D.A., García–Viguera, C., (2007). Effects of microwave cooking conditions on bioactive compounds present in broccoli inflorescences. *J. Agric. Food Chem.* 55: 10001–10007.

Lucock, M., Yates, Z., (2005). Folic acid–vitamin and panacea or genetic time bomb? *Nat. Rev. Genet.* 6: 235–240.

Lule, S.U., Xia, W., (2005). Food phenolics, pros and cons: A review. *Food Rev. Int.* 21: 367–388.

Maiani, G., Periago–Castón, M.J., Catasta, G., Toti, E., Goñi–Cambrodón, I., Bysted, A., Granado–Lorencio, F., Olmedilla–Alonso, B., Knuthsen, P., Valoti, M., Böhm, V., Mayer–Miebach, E., Behsnilian, D., Schlemmer, U., (2009). Carotenoids: Actual knowledge on food sources, intakes, stability and bioavailability and their protective role in humans. *Mol. Nutr. Food Res.* 53: 194–218.

Marks, H.S., Hilson, J. A., Leichtweis, H.C., Stoewsand, G.S., (1992). S–Methylcysteine sulfoxide in *Brassica* vegetables and formation of methyl methanethiosulfinate from Brussels sprouts. *J. Agric. Food Chem.*40: 2098–2101.

Martínez–Hernández, G.B., Gómez, P., Pradas, I., Artés, F., Artés–Hernández, F., (2011). Moderate UV–C pretreatment as a quality enhancement tool in fresh–cut Bimi® broccoli. *Postharvest Biol. Technol.* 62: 327–337.

Martínez–Hernández, G.B., Artés–Hernández, F., Gómez, P., Artés, F., (2012a). Quality changes after vacuum–based and conventional industrial cooking of kailan–hybrid broccoli throughout retail cold storage. *Lebensm. Wiss. Technol.* 50: 707–714.

Martínez–Hernández, G.B., Artés–Hernández, F., Colares–Souza, F., Gómez, P., García–Gómez, P., Artés, F., (2012b). Innovative cooking techniques for improving the overall quality of a kailan–hybrid broccoli. *Food Bioprocess Tech.* In press. DOI: 10.1007/s11947–012–0871–0.

Martínez–Hernández, G.B., Artés–Hernández, F., Gómez, P., Artés, F., (2013a). Comparative behaviour between kailan–hybrid and conventional fresh–cut broccoli throughout shelf–life. *Lebensm. Wiss. Technol.* 50: 298–305.

Martínez–Hernández, G.B., Artés–Hernández, F., Gómez, P., Formica, A.C., Artés, F., (2013b). Combination of electrolysed water, UV–C and superatmospheric O_2 packaging for improving fresh–cut broccoli quality. *Postharvest Biol. Technol.* 76: 125–134.

Martínez–Hernández, G.B., Artés–Hernández, F., Gómez, P., Artés, F., (2013c). Induced changes in bioactive compounds of kailan–hybrid broccoli after innovative processing and storage. *J. Funct. Foods*. In press. DOI: 10.1016/j.jff.2012.09.004.

Maruyama, F.T., (1970). Identification of dimethyl disulphide as a major aroma component of cooked brassicaceous vegetables. *J. Food Sci.* 35: 540–543.

Mathews–Roth, M.M., (1990). Plasma concentrations of carotenoids after large doses of beta–carotene. *Am. J. Clin. Nutr.* 52: 500–501.

McClure, J.W., (1975). Physiological functions of flavonoids. In: Harborne, J.B., Mabry, T.J., Mabry, H. (Eds.). The flavonoids. Chapman, London, UK. pp. 970–1055.

McKillop, D. McNulty, H. Scott, J.. McPartlin, J Strain, J. Bradbury I., Girvan, J., Hoey, L., McCreedy, R., Alexander, J., Patterson, B.K., Hannon–Fletcher, M., Pentieva, K., (2006). The rate of intestinal absorption of natural food folates is not related to the extent of folate conjugation. *Am. J. Clin. Nutr.* 84: 167–173.

Mithen, R., Faulkner, K., Magrath, R., Rose, P., Williamson, G., Marquez, J., (2003). Development of isothiocyanate–enriched broccoli, and its enhanced ability to induce phase 2 detoxification enzymes in mammalian cells. *Theor. Appl. Genet.* 106: 727–734.

Moreno, D.A., Carvajal, M., López–Berenguer, C., García–Viguera, C., (2006). Chemical and biological characterisation of nutraceutical compounds of broccoli. *J. Pharm. Biom. Anal.* 41: 1508–1522.

Munday, R., Munday, C.M., (2004). Induction of Phase II detoxification enzymes in rats by plant–derived isothiocyanates: comparison of allyl isothiocyanate with sulforaphane and related compounds. *J. Agric. Food Chem.* 52: 1867–1871.

Murkovic, M., Gams, K., Draxl, S., Pfannhauser, W., (2000). Development of an Austrian carotenoid database. *J. Food Comp. Anal.* 13: 435–440.

Musk, S.R., Smith, T.K., Johnson, I.T., (1995). On the cytotoxicity and genotoxicity of allyl and phenethyl isothiocyanates and their parent glucosinolates sinigrin and gluconasturtiin. *Mutat. Res.* 34: 19–23.

Nagaharu U., (1935). Genome analysis in *Brassica* with special reference to the experimental formation of *B. napus* and peculiar mode of fertilization. *Japan. J. Bot.* 7: 389–452.

Nishino, H., (1998). Cancer prevention by carotenoids. *Mutat. Res.* 402: 159–163.

Noichinda, S., Bodhipadma, K., Mahamontri, C., Narongruk, T., Ketsa, S., (2007). Light during storage prevents loss of ascorbic acid, and increases glucose and fructose levels in Chinese kale (*Brassica oleracea* var. Alboglabra). *Postharvest Biol. Technol.* 44: 312–315.

Padayatty, S.J., Levine, M., (2001). New insights into the physiology and pharmacology of vitamin C. *CMAJ.* 164: 353–355.

Page, T., Griffiths, G., Buchanan–Wollaston, B., (2001). Molecular and biochemical characterization of postharvest senescence in broccoli. *Plant Physiol.* 125: 718–727.

Perry, J., (1971). Folate analogues in normal mixed diets. *Br. J. Haematol.* 21: 435–441.

Redovniković, I.R., Repajić, M., Fabek, S., Delonga, K., Toth, N., Furač, J.V., (2012). Comparison of selected bioactive compounds and antioxidative capacity in different broccoli cultivars. *Acta Aliment. Hung.* 41: 221–232.

Redovniković, I.R., Repajić, M., Fabek, S., Delonga, K., Toth, N., Furač J.V., (2012). Comparison of selected bioactive compounds and antioxidative capacity in different broccoli cultivars. *Acta Alimentaria.* 41: 221–232.

Ribaya–Mercado, J.D., Blumberg, J.B., (2004). Lutein and zeaxanthin and their potential roles in disease prevention. *J. Am. Coll. Nutr.* 23: 567–587.

Rosa, E.A.S., Haneklaus, S.H., Schnug, E., (2002). Mineral content of primary and secondary inflorescences of eleven broccoli cultivars grown in early and late seasons. *J. Plant Nutr.* 25: 1741–1751.

Rosa, E.A.S., Heaney, R.K., Fenwick, G.R., Portas, C.A.M., (1997). Glucosinolates in crop plants. *Hortic. Rev.* 19: 99–215.

Rosa, E.A.S., Heany, R.K., Portas, C.A.M., Fenwick, G.R., (1996). Changes in glucosinolate concentrations in *Brassica* crops (*B. oleracea* and *B. napus*) throughout growing seasons. *J. Sci. Food Agric.* 71: 237–244.

Rosa, E.A.S., Rodrigues, A.S., (2001). Total and individual glucosinolate content in 11 broccoli cultivars grown in early and late seasons. *Hort. Sci*. 36: 56–59.

Rossiter, R.C., Barrow, N.J., (1972). Physiological and ecological studies on the oestrogenic isoflavones in subterranean clover (*T. subterraneum* L.). IX. Effects of sulphur supply. *Aust. J. Agric. Res*. 23:411–418.

Sanderson, P., McNulty, H., Mastroiacovo, P., McDowell, I.F.W., Melse–Boonstra, A., Finglas, P.M., Gregory, J.F., (2003). Folate bioavailability: UK Food Standards Agency workshop report. *Br. J. Nutr.* 90: 473–479.

Schonhof, I., Krumbein, A., Brückner, B., (2004). Genotypic effects on glucosinolates and sensory properties of broccoli and cauliflower. *Nahrung/Food*. 48: 25–33.

Schonhof, I., Krumbein, A., Schreiner, M., Gutezeit, B., (1999). Bioactive substances in cruciferous products. In: Hägg, M., Ahvenainen, R., Evers, A.M., Tiilikkala, K. (Eds.). Agri–Food II – Quality management of fruits and vegetables. The Royal Society of Chemistry, Cambridge, UK. 222–226.

Scott, J.M., Weir, D.G., (1976). Folate composition, synthesis and function in natural materials. *Clin. Haematol.* 5: 547–568.

Shahidi, F., Naczk, M., (2004). Phenolics in food and nutraceuticals. CRC Press, Boca Raton FL, USA. 1–2.

Simopoulos, A.P., (1999). Essential fatty acids in health and chronic disease. *Am. J. Clin. Nutr.* 70: 560–569.

Smith, T.K., Lund, E.K., Musk, S.R.R., Johnson, I.T., (1998). Inhibition of DMH–induced aberrant crypt foci, and induction of apoptosis in rat colon, following oral administration of a naturally occurring glucosinolate. *Carcinogenesis.* 19: 267–273.

Snodderly, D.M., (1995). Evidence for protection against age–related macular degeneration by carotenoids and antioxidant vitamins. *Am. J. Clin. Nutr.* 62: 1448–1461.

Soengas, P., Sotelo, T., Velasco, P., Cartea, M.E., (2011). Antioxidant properties of *Brassica* vegetables. *Funct. Plant Sci. Biotechnol*. 5: 43–55.

Soetan, K.O., Olaiya, C.O., Oyewole, O.E., (2010). The importance of mineral elements for humans, domestic animals and plants: A review. *Afr. J. Food Sci*. 4: 200–222.

Souci, W., Fachmann, W., Kraut, H., 2000. Food composition and nutrition tables. 6th edition. Medpharm Scientific Publishers CRC Press, London, UK. 697.

Talalay, P., Fahey, J.W., Holtzclaw, W.D., Prestera, T., Zhang, Y., (1995). Chemoprotection against cancer by Phase II enzyme induction. *Toxicol. Lett.* 82: 173–179.

Tanimura & Antle, 2011. Accesed on October 2012. http://www.taproduce.com/trade/fresh–vegetables–detail.php?id=15

Tawashi, R., Cousineau, M., Denis, G., (1982). Crystallisation of calcium oxalate dihydrate in normal urine in presence of sodium copper chlorophyllin. *Urol. Res.* 10: 173–176.

Tonguç¸ M., Griffiths, P.D., (2004). Development of black rot resistant interspecific hybrids between *Brassica oleracea* L. cultivars and *Brassica* accession A 19182, using embryo rescue. *Euphytica*. 136: 313–318.

Traka, M., Gasper, A.V., Melchini, A., Bacon, J.R., Needs, P.W., Frost, V., Chantry, A., Jones, A.M.E., Ortori, C.A., Barrett, D.A., Ball, R.Y., Mills, R.D., Mithen, R.F., (2008). Broccoli consumption interacts with GSTM1 to perturb oncogenic signalling pathways in the prostate. *Pros One.* 3 (7), e2568.

Traka, M., Mithen, R., (2009). Glucosinolates, isothiocyanates and human health. *Phytochem. Rev*. 8:269–282.

Ulrich, D., Krumbein, A., Schonhof, I., Hoberg, E., (1998). Comparison of two sample preparation techniques for sniffing experiments with broccoli (*Brassica oleracea* var. *italic* Plenck). *Food/ Nahrung*. 42: 392–394.

Vallejo, F., Tomás–Barberán, F.A., García–Viguera, C., (2002). Potential bioactive compounds in health promotion from broccoli cultivars grown in Spain. *J. Sci. Food Agric.* 82: 1293–1297.

Vallejo, F., Tomás–Barberán, F.A., García–Viguera, C., (2003). Effect of climatic and sulphur fertilisation conditions, on phenolic compounds and vitamin C, in the inflorescences of eight broccoli cultivars. *Eur. Food Res. Technol.* 216: 395–401.

Van Doorn, H.E., Van der Kruk, G.C., Van Holst, G.–J., Raaijmakers–Ruijs, C.M.E., Postma, E., Groeneweg, B., Jongen, W.H.F., (1998). The glucosinolates sinigrin and progoitrin are important determinants for taste preference and bitterness of Brussels sprouts. *J. Sci. Food Agric.* 78: 30–38.

Van Etten, C.H., Daxenbichler, M.E., Williams, P.H., Kwolek, W.F., (1976). Glucosinolates and derived products in cruciferous vegetables. Analysis of the edible part from 22 varieties of cabbage. *J. Agric. Food Chem.* 24: 452–455.

Wallsgrove R.M., Bennett R.N., (1995). The biosynthesis of glucosinolatos in *Brassicas*. In: Wallsgrove, R.M. (Ed.). Amino Acids and their derivatives in higher plants. Cambridge University Press, Cambridge, UK.

Walters, D.E., (1996). How are bitter and sweet tastes related? *Trends Food Technol.* 7: 399–403.

Walters, D.E., Roy, G., (1996). Taste Interactions of Sweet and Bitter Compounds. In: McGorrin, R.J., Leland, J.V. (Eds.). Flavour–food Interactions. ACS Symposium Series 633, Am. Chem. Soc., Washington DC, USA. 130–142.

Wu, X., Zhou, Q., Xu, K., (2009). Are isothiocyanates potential anti–cancer drugs? *Acta Pharmacol. Sin.* 30: 501–512.

Zhang, Y., Talalay, P., Cho, C.G., Posner, G.H., (1992). A major inducer of anticarcinogenic protective enzymes from broccoli: Isolation and elucidation of structure. *Proc. Natl. Acad. Sci. U.S.A.* 89: 2399–2403.

Zhuang, H., Hildebrand, D.F., Barth, M.M., (1997). Temperature influenced lipid peroxidation and deterioration in broccoli buds during postharvest storage. *Postharvest Biol. Technol.* 10: 49–58.

In: Brassicaceae
Editor: Minglin Lang

ISBN: 978-1-62808-856-4

Chapter 4

DEGRADATION OF CHLOROPHYLL DURING POSTHARVEST SENESCENCE OF BROCCOLI

***Gustavo A. Martínez*[*1,2], *Pedro M. Civello*[2,3] and *María E. Gómez-Lobato*[2]**

[1]Instituto de Investigaciones Biotecnológicas - Instituto Tecnológico de Chascomús (IIB-INTECH) UNSAM-CONICET, Buenos Aires, Argentina
[2]Instituto de Fisiología Vegetal (INFIVE) UNLP-CONICET, La Plata, Argentina
[3]Facultad de Ciencias Exactas, Universidad Nacional de La Plata (UNLP), La Plata, Argentina

ABSTRACT

Broccoli (*Brassica oleracea* L. *Italica*) is a floral vegetable rich in diverse compounds such as vitamins A and C, antioxidants, and anti-carcinogenic compounds. Floral heads of broccoli are composed of hundreds of florets arranged in whorls on top of stems. For consumption, they are harvested in an immature stage when male and female reproductive structures are still surrounded by petals and enclosed by chlorophyll-containing sepals. Harvesting causes heads to experience disruption of energy, nutrition, and hormone supplies, thus causing fast senescence and chlorophyll degradation in sepals. Catabolism of chlorophyll leads to yellowing, which is the main sign of quality deterioration in harvested broccoli.

In recent years, a pathway of chlorophyll degradation that is active during senescence has been elucidated. Most of the genes and enzymes of this pathway have been characterized in *Arabidopsis thaliana*, although many of them have their orthologs in broccoli. In chloroplasts, chlorophyll molecules interact with several proteins forming light-harvesting complexes, which must be destabilized as a prerequisite for the subsequent degradation of chlorophyll. It has been described that a protein named SGR interacts with light-harvesting complex II, enhancing destabilization of these chlorophyll-apoprotein complexes. After that, phytol is hydrolyzed by the action of chlorophyllase or pheophytinase, and Mg^{2+} is removed by a metal-chelating substance. Then, the porphyrin ring of the pheophorbide is oxygenolytically opened by pheophorbide a oxygenase. The

[*] Corresponding author.

product of this reaction is red chlorophyll catabolite, which is site-specifically reduced by red chlorophyll catabolite reductase to yield the primary fluorescent chlorophyll catabolite, a product that is exported to vacuoles.

Several attempts have been made to extend broccoli postharvest life, mainly by reducing the senescence rate and the loss of green color. To that end, refrigerated storage, controlled and modified atmospheres, heat treatments, UV applications, 1-methylcyclopropene and ethanol have been used. In this review, research on changes in enzyme activity and expression of genes, associated with chlorophyll catabolism during postharvest senescence of broccoli is reviewed. Furthermore, the effect of different hormonal and postharvest physical treatments on the expression of the mentioned genes and enzymes is examined.

Keywords: Broccoli, postharvest, senescence, chlorophyll catabolism

INTRODUCTION

Broccoli Characteristics

Broccoli (*Brassica oleracea* L. var *Italica*) is a vegetable with a high nutritional value. Fresh broccoli heads have low caloric levels, high content of fibers, and high levels of vitamins A and C. Besides, broccoli possesses a wide range of phytochemicals, including phenolic compounds, carotenoids and glucosinolates (Jeffery and Araya, 2009; Singh *et al.*, 2007; Wold *et al.*, 2006). The high content of ascorbic acid and phenolic compounds gives this vegetable an important antioxidant capacity. Broccoli is also considered a functional food due to its high levels of glucosinolates. These compounds are chemically stable until they come into contact with the enzyme myrosinase (ß-tioglucoside glucohydrolase, EC 3.2.3.1), which is stored in cellular compartments separated from glucosinolates in healthy tissues. The physiological role of these compounds is the defense against herbivores. When the tissue is disrupted, glucosinolates are liberated from vacuoles and interact with myrosinase. The enzyme catalyzes the hydrolysis of the compound, releasing a sugar and rendering an aglycon unstable that rapidly decomposes into isothiocyanates, thiocyanates and/or nitriles, substances that act as insect repellent (Halkier and Gershenzon, 2006). Nevertheless, from the point of view of human health, it has been demonstrated that isothiocyanates produced by the decomposition of glucosinolates have a protective effect against colon, bladder and lung cancer (Jeffery and Araya, 2009), and that the regular consumption of broccoli and other crucifers with high content of glucosinolates diminishes the risk of developing this type of pathologies (Jeffery and Araya, 2009). For these reasons, many investigators started to refer to broccoli as the "crown jewel of nutrition."

Inflorescences or heads of broccoli for consumption are harvested during development. Heads are composed of several florets arranged in whorls held in place by a fleshy stem. Individual florets contain hundreds of male and female reproductive structures surrounded by immature petals and enclosed with chlorophyll-containing sepals. As any organ under development, inflorescences need abundant supply of water, nutrients and hormones. Harvesting interrupts this flow and causes a severe stress, which in turn induces an early beginning of senescence. During development, the quality and characteristics of inflorescences will depend on the rate of absorption of nutrients, photosynthesis, transpiration, respiration and other metabolic processes of the whole plant. Once harvested, the quality of heads depends on the internal rates of transpiration and respiration. Precisely, broccoli heads possess a high respiration rate, which accelerates the process of senescence and deterioration. Heads of broccoli have a shelf life of approximately 3-4 days at 20 °C and 3-4 weeks at 0 °C.

Senescence is a highly regulated process, in which some metabolic routes are induced while others are silenced, as a consequence of activation and deactivation of diverse genes. The genetic processes combine several signaling factors with the expression of many genes in order to disrupt cellular architecture and macromolecules. Metabolism changes from autotroph to heterotroph and, during senescence, routes of recovery of materials are activated. In addition, cells also perform a controlled differentiation of structures. The most visible and relevant of which are those found in chloroplasts, whose thylakoids are dismantled and the chloroplasts transformed into gerontoplasts (Hörtensteiner, 2006; Matile *et al.*, 1999). Main studies on senescence were conducted in leaves, but Page *et al.* (2001) demonstrated that senescence of sepals of broccoli inflorescences share characteristics with senescence of leaves.

The main purpose of senescence is to mobilize and to recycle nutrients; for which macromolecules are degraded. Most proteins are located inside the chloroplast, and numerous studies have revealed a decrease in the content of stromal proteins, such as RUBISCO; and thylakoidal proteins, such as proteins of the LHCPII, during senescence. This degradation is accompanied by an increase in the activity of proteases, both chloroplastic and vacuolar. In broccoli, several studies have shown a decrement in the total and soluble proteins (Büchert *et al.*, 2011b; Lemoine *et al.*, 2009), and a simultaneous increment in the activity and expression of genes encoding proteases (Wang *et al.*, 2004) during postharvest senescence. Degradation of proteins drives to the synthesis of nitrogenous amino acids (glutamine and asparagine), which are translocated to other organs of the plant. Nevertheless, in broccoli and other vegetables that have been harvested, the N cannot be translocated and it is either accumulated as NH_4^+, or eliminated as NH_3 (Downs *et al.*, 1997a; King and Morris, 1994).

During senescence, an important catabolism of lipids also takes place. Thylakoidal membranes are an important source of lipids that must be mobilized to be used as source of carbon and energy by the way of ß-oxidation and the glyoxylate cycle, metabolic routes that are important when the contribution of sugars is scarce (Graham and Eastmond, 2002). Genes that encoded for enzymes like phospholipases, phosphatidic acid phosphatase and lipoxygenases have an increased expression during senescence (Thompson *et al.*, 1998). Lipid catabolism contributes to the energy supply to the cell but also alters the structure and functionality of membranes. As a consequence, in advanced stages of senescence, cellular compartmentalization is lost, contributing to the loss of tissue integrity and to the dehydration. In broccoli, an important degradation of lipids has been detected (Zhuang *et al.*, 1997) and an increase in the activity and expression of enzymes and genes related to this catabolism has been described (Page). Thus, diverse studies have shown an increase in the loss of electrolytes during the senescence of broccoli (Duarte-Sierra *et al.*, 2012).

Senescence is associated to several metabolic changes that require energy provided by sugars. Harvesting interrupts the input of nutrients from the plant and triggers senescence, which in turn interrupts the photosynthesis in sepals. Both facts considerably change the supply of sugars necessary to maintain tissue integrity. In broccoli, diverse researches have shown to decrement in the level of soluble sugars during postharvest (Finger *et al.*, 1999;

King and Morris, 1994). Besides, considering that metabolism of sepals is similar to that of leaves, an increment in the content of starch during the day was detected (Hasperué *et al.*, 2011), but a marked decrement during the postharvest was identified, as well (Finger *et al.*, 1999).

As in other green tissues, senescence of broccoli is regulated by cytokinins and ethylene, which play antagonistic roles. In intact plants, cytokinins are synthesized in the roots and transported to the flowers, and their concentration diminishes abruptly after harvest, and this is generally considered to induce organ senescence (Tian *et al.*, 1994). The treatment with cytokinins delays the physiological changes that usually_accompany the senescence of florets (Downs *et al.*, 1997b). Moreover, induction of hydric stress at the pre-harvest stage during plant growth induces cytokinins biosynthesis, which in turn delays senescence during postharvest (Zaicovski *et al.*, 2008). On the contrary; ethylene seems to be the principal promoter of senescence and yellowing. The level of 1-amino-cyclopropane-1-carboxylic acid increases during broccoli postharvest (Pogson *et al.*, 1995), and the incubation of broccoli heads in ethylene-enriched atmospheres enhances the senescence (King and Morris, 1994). Treatments with sugars, which reduce ethylene sensibility (Nishikawa *et al.*, 2005), or with 1-methylcyclopropene (1-MCP), an irreversible blocking agent of ethylene receptor (Ku and Wills, 1999), delayed postharvest senescence of broccoli.

The most evident symptom of senescence in green tissues is the gradual loss of green color over time, caused by the degradation of chlorophyll. It is estimated that about 1.2×10^9 tons of chlorophyll are degraded annually worldwide. Most of this process occurs in the leaves of deciduous trees, although there is also chlorophyll degradation during fruit ripening, senescence of various organs, and normal turnover of no senescent organs. The chlorophyll degradation pathway is detailed in the following section.

Chlorophyll Catabolic Pathway

Chlorophyll degradation begins with disruption of pigment-protein complexes inside of chloroplasts, a process that frees chlorophyll molecules. A protein without enzymatic activity (SGR) interacts with protein thylakoidal complexes, destabilizing their structure, and contributing to the release of chlorophyll; as a prerequisite for the subsequent degradation of both apoprotein and chlorophyll (Hörtensteiner, 2009). Chlorophyll is a photoactive compound that generates free radicals that can damage several cellular components. For this reason chlorophyll must be quickly degraded as a detoxification mechanism (Matile *et al.*, 1999).

The currently accepted biochemical pathway of degradation of chlorophyll comprises two stages, which are divided according to the moment of the tetrapyrrole ring opening (Figure 1). Products of the first stage (prior to the rupture of the macrocycle) totally or partially retain the green color, while those of the second stage lose that coloration and are almost colorless. The first stage includes modifications to the side chain of the macrocycle, the hydrolysis of phytol, release of Mg^{2+} from the tetrapyrrole, and other reactions that can vary among species. The second stage is essential to the loss of green color characteristic during senescence. In most of the cases studied, no degradation intermediates were accumulated at a detectable amount, which suggests that a number of catabolic reactions occurred in coordination.

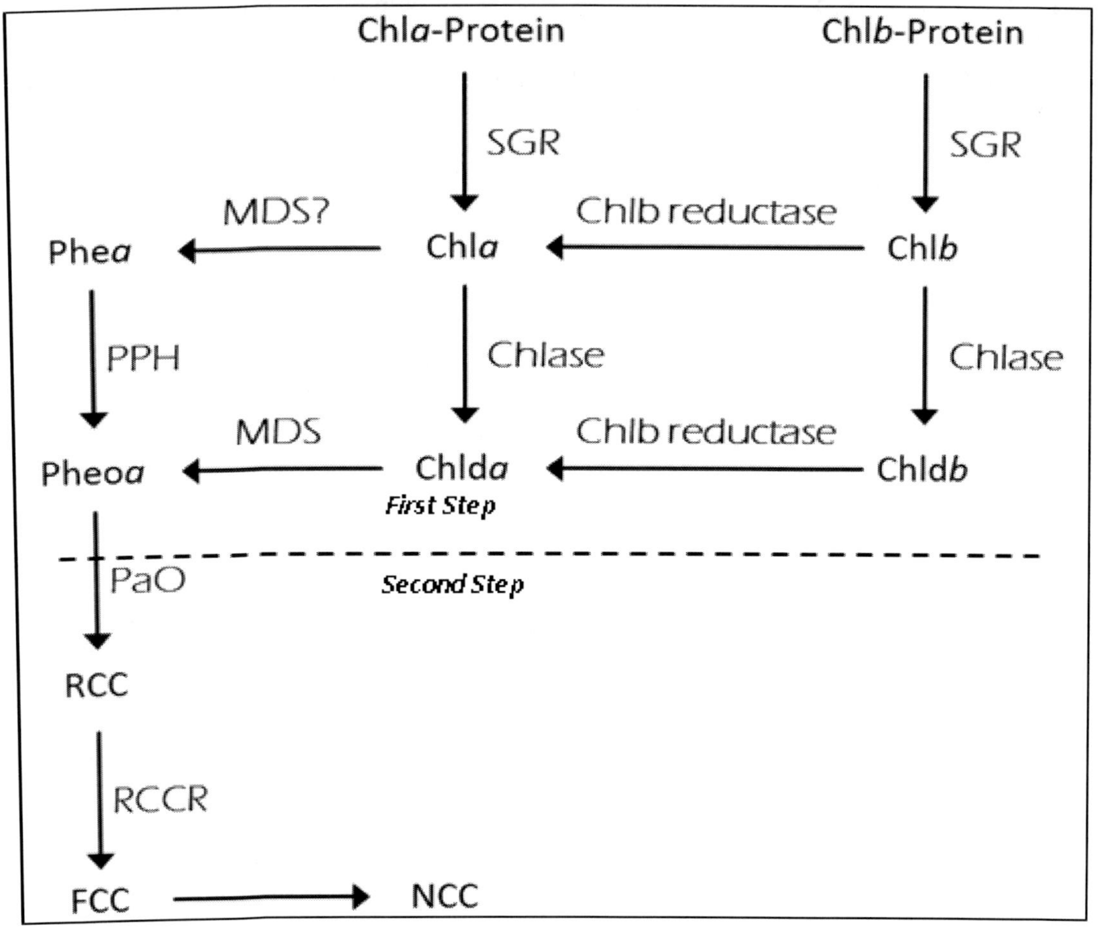

Figure 1. Pathway of chlorophyll degradation. First Step: before PaO; Second Step: after PaO. Chl*a*: chlorophyll a; Chl*b*: chlorophyll b; Chld*a*: chlorophyllide a; Chld*b*: chlorophyllide b; Phe*a*: pheophytin a; Pheo*a*: pheophorbide a; RCC: red chlorophyll catabolite; FCC: fluorescent chlorophyll catabolite; NCC: non-fluorescent chlorophyll catabolite. SGR: stay green; Chlase: chlorophyllase; MDS: magnesium dechelatase; PaO: pheophorbide a oxygenase; RCCR: red chlorophyll catabolite reductase.

Once chlorophyll molecules are released, the first catabolic reaction that occurs is the elimination of phytol. Chlorophyllase (chlorophyll-chlorophyllido hydrolase, CLH, EC 3.1.1.14) was the first enzyme to be identified as causing this reaction (Hörtensteiner, 2006; Lee *et al.*, 2010; Schenk *et al.*, 2007; Tsuchiya T *et al.*, 1999). In many cases, this enzyme is a glycosylated protein associated to hydrophobic chloroplast membranes and other organelles, and is characterized by its functional latency. Although at the outset it was believed that chlorophyllases were located in the chloroplast membrane (Brandis *et al.*, 1996; Matile *et al.*, 1987) no transmembrane domains have been found in the sequences obtained to date, according to the hydropathy profiles. This suggests that chlorophyllases are not intrinsic membrane proteins. The functional property of latency observed in CLH could be simply due to the spatial separation between the enzyme and its substrate. Chlorophyllase catalyzes the hydrolysis of the ester linkage between the chlorophyll and phytol, reaction that is considered as the first step in the catabolism of chlorophyll. The products of such reaction are phytol and chlorophyllide ((Benedetti and Arruda, 2002; Matile *et al.*, 1999; Takamiya *et al.*, 2000).

Recently, the true involvement of chlorophyllases in chlorophyll breakdown has been questioned, given that not all isolated genes have a chloroplast transit peptide, suggesting alternative pathways occurring outside of the chloroplast or involvement of enzymes, other than CLH (Hörtensteiner, 2006; Takamiya *et al.*, 2000). Schenk *et al.* (2007) have shown that Arabidopsis mutants with interrupted expression for both known chlorophyllases are still able to degrade chlorophyll during senescence, indicating that these genes are not essential for this catabolic process. Based on these findings, Schelbert *et al.* (2009) set out to reveal the true CLH responsible for chlorophyll dephytilation in Arabidopsis. Instead, their findings revealed the existence of a new enzyme, termed pheophytinase (PPH), which would act as a pheophytin hydrolase.

Assuming that chlorophyllase is the first enzyme in the catabolic pathway, the next step would be the elimination of the central ion Mg^{2+} from chlorophyllide. With regard to this step, studies have revealed conflicting data, and for this reason, consensus has not been reached. Two types of activity were found: one associated with a low molecular weight compound stable at high temperatures (Shioi *et al.*, 1996), and other associated with a thermally labile protein that is probably part of chloroplast membranes (Vicentini *et al.*, 1995). The important difference between these two proposals regarding activity can be explained considering that the low molecular mass compound is a cofactor of an enzyme of higher molecular mass (Costa *et al.*, 2002). More recently, the development of new proposals has suggested that protein with Mg-dechelatase activity can only act *in vitro* on a widely used artificial substrate, chlorophyllin, but not on the *in vivo* substrate, clorophyllide, while the low molecular weight compound acts on both substrates (Kunieda *et al.*, 2005; Suzuki *et al.*, 2005; Suzuki and Shioi, 2002). However, if chlorophyllases are not the main enzymes that release phytol, and dephytilation occurs on Mg-free chlorophyll, then the whole degradation pathway must be revised, especially with regard the early reactions. In the new suggested model, Mg release seems to precede phytol cleavage, producing pheophytin, which is then dephytilated by PPH to give pheophorbide (Schelbert *et al.*, 2009).

Besides chlorophyllase and Mg-dechelatase, other modifications of chlorophyll before ring opening were described. An enzyme known as pheophorbidase was described and purified from *Chenopodium album*. This enzyme catalyzes the hydrolysis of methyl ester bond of pheophorbide isocyclic ring. The product of this reaction is not stable; hence it is converted to pyropheophorbide nonenzymatically. An interesting aspect is that pheophorbidase is located outside the chloroplast, indicating that if pheophorbide a is the true substrate of this enzyme, there would be a degradation pathway, the initial steps of which occur outside the chloroplast (Takamiya *et al.*, 2000).

The stage of tetrapyrrole ring opening is crucial for the loss of green color in the tissues and for eliminating the photo-activity of chlorophyll. This reaction is carried out by the enzyme pheophorbide a oxygenase (PAO), which is a component of the inner membrane of gerontoplasts and chromoplasts. PAO catalyzes the oxygenolitic breakdown of the ring between the C4 and C5 by adding two oxygen atoms and four hydrogen atoms. The product of this reaction is the first identifiable colorless compound, RCC (red chlorophyll catabolite). PAO is a Rieske-type iron-sulfur oxygenase that requires the presence of reduced ferredoxin and NADPH and both compounds are present in the gerontoplasts, but not in presenescent tissues (Hörstensteiner *et al.*, 1998; Pruzinskà *et al.*, 2003; Takamiya *et al.*, 2000). For this reason, it was initially believed that its activity was limited to senescence. However, recent research has suggested that PAO activity is present from before the onset of senescence

(Pruzinskà *et al.*, 2003; Pruzinská *et al.*, 2005; Roca *et al.*, 2004). The enzyme also has a high specificity towards the pheophorbide a, while the pheophorbide b induces a competitive inhibition (Engel *et al.*, 1996; Iturraspe *et al.*, 1994).

A reduction of the RCC methine bridge occurs coupled with the previously described oxygenase reaction. Such reaction is catalyzed by the enzyme RCC reductase, which is a soluble enzyme localized in the stroma of chloroplasts (Rodoni *et al.*, 1997). This reaction produces a fluorescent colorless compound called pFCC (by primary fluorescent chlorophyll catabolite). Such reaction requires the presence of ferredoxin and the absence of oxygen, indicating that reactions of RCC and PAO are very closely related; due to the fact that the oxygenase reaction consumes oxygen and maintains anaerobic microenvironment for the reductase (Takamiya *et al.*, 2000; Wüthrich *et al.*, 2000).

While the chlorophyll degradation pathway seems to be very similar to the formation of pFCC in different species, due to the wide variety found in these compounds, it may be assumed that a series of reactions occurs after formation. The pFCC are converted to fluorescent chlorophyll catabolites (FCC) with several modifications that vary among species, such as demethylation and hydroxylation (Hörtensteiner, 2006). Most of the enzymes involved in these steps have not yet been identified, although such reactions_are known to cause an increase in the solubility of catabolites. Recently, a member of the methylesterase protein family (MES16) has been identified described in Arabidopsis. It specifically demethylates chlorophyll catabolites at the level of FCCs (Christ *et al.*, 2011). It is considered that the modified FCCs are transported to the vacuole by ATP dependent translocators on the tonoplast, and then converted to NCC (nonfluorescent chlorophyll catabolites) through a rearrangement of the double bond tetrapyrrole and its adjacent methine bridge.

Modifications to the NCC are diverse and very much depend on the species. Furthermore, it is not entirely clear whether NCCs are stored in vacuoles without further degradation. In some species accumulated NCC levels correspond to the total degraded chlorophyll, indicating that the NCC is the end of chlorophyll degradation (Hörstensteiner *et al.*, 1998; Matile *et al.*, 1999). However, in other species, such as tobacco and spinach, NCC concentrations are higher in early stages of leaf senescence, but lower in later stages (Hörtensteiner, 2006).

Chlorophyll b is an accessory pigment in light harvesting complexes that can represent up to 30% of the total chlorophyll. Despite its relative abundance in higher plants, chlorophyll catabolites derive from chlorophyll a. This indicates that there should be a conversion of chlorophyll b to chlorophyll a early in the degradation pathway, which is supported by several observations. For example, enzyme pheophorbide a oxygenase only takes pheophorbide a as substrate, while the pheophorbide b is a competitive inhibitor. The inhibition of pheophorbide a oxygenase causes the accumulation of both forms of chlorophyllide, but only of pheophorbide a. These findings suggest the presence of enzymes with chlorophyll b reductase activities. Recently, it has been described that the conversion of chlorophyll b to chlorophyll a requires two steps. Horie *et al.* (2009) showed that a gene named *AtNYC* encodes for a chlorophyll or chlorophyllide b reductase transforming chlorophyll b to 7-hydroxymethyl chlorophyll a. This compound is finally transformed by 7-hydroxymethyl chlorophyll a reductase producing chlorophyll a (Meguro *et al.*, 2011).

All genes encoding the enzymes described above, except Mg-dechelatasa, have been cloned and characterized, particularly in *Arabidopsis thaliana* (Hörtensteiner and Kräutler, 2011).

Additionally, an alternative route of chlorophyll catabolism has been also suggested. Studies conducted on different systems support the possible participation of peroxidase in the catabolism of chlorophyll (Maeda *et al.*, 1998; Martínez *et al.*, 2001; Funamoto *et al.*, 2002; Funamoto *et al.*, 2003; Costa *et al.*, 2004; Funamoto *et al.*, 2006).

Chlorophyll Catabolism in Broccoli

At harvest, broccoli inflorescences are in development; the sepals are closed and surround the floral structures. Sepals have a deep green color due to a high concentration of chlorophyll. The stress caused by harvest triggers senescence, which is manifested by an intense degreening and yellowing. At 20 °C, broccoli almost completely loses their chlorophyll molecules in about 3-4 days (Deschene *et al.*, 1991; Finger *et al.*, 1999; Zapata *et al.*, 2012). Several studies have shown an increment in chlorophyll catabolites content during broccoli senescence, mainly those of the first steps of the catabolic pathway. Kaewsuksaeng *et al.* (2006) showed that levels of chlorophyllide a, pheophorbide a, pyropheophorbide a, $C13^2$-hydroxychlorophyll and pheophytin a increased when broccoli heads are stored at 15 °C. Moreover, content of chlorophyllide a, $C13^2$-hydroxychlorophyll a and pheophytin a decreased concomitantly with the increment of pheophorbide a levels (Kaewsuksaeng *et al.*, 2006). So far, catabolites, other than pheophorbide a, (such as NCCs) have not been identified in broccoli.

High enzymatic activity and expression of the related encoding genes involved in the early stages of chlorophyll degradation were detected in broccoli. Chlorophyllase is one of the most extensively studied enzimes, and conflicting findings have been revealed in this respect: Funamoto *et al.* (2002) found no changes in chlorophyllase activity during senescence; whereas Costa *et al.* (2004) found an increment in the activity during the same period, which was regulated by ethylene and cytokinin hormones. Three genes encoding chlorophyllases were described (*BoCHL1*; *BoCHL2* and *BoCHL3*), but their expression during senescence is even more intriguing. Büchert *et al.* (2011) showed that expression of *BoCHL1* is negatively regulated during senescence whereas *BoCHL2* expression is enhanced during the same period. Differently, Aiamla-or *et al.* (2012) found that three genes have a decreased expression during senescence. These results would suggest a minor or null role of chlorophyllase in chlorophyll catabolism as it was previously showed for Arabidopsis (Hörtensteiner *et al.*, 2009; Schelbert *et al.*, 2009). However, transgenic broccoli with antisense-suppression of *BoCLH1* showed a delayed postharvest yellowing of heads and leaves (Chen *et al.*, 2008).

In relation to Mg-dechelatase, several studies have described an increment of activity during broccoli senescence (Kaewsuksaeng *et al.*, 2010; Takahashi *et al.*, 2001) and also a regulation by ethylene and cytokinins (Costa *et al.*, 2004). However, as in any other plant system, much work remains to be done to purify the enzyme and/or clone the related gene. The traditional route of chlorophyll degradation supposes the release of Mg^{2+} from chlorophyllide to form pheophorbide. However, in broccoli, accumulation of pheophytins has been detected during postharvest senescence (Kaewsuksaeng *et al.*, 2006; Costa *et al.*, 2006b), suggesting the presence of an unknown mechanism for the release of Mg^{2+} from chlorophyll and providing a substrate for pheophytinase. The possibility for Mg-dechelatase to act directly on chlorophyll must be re-examined, although the data of *in vitro* activity do not support this fact (Ni *et al.*, 2001). The alternative route of chlorophyll degradation recently proposed (Schelbert *et al.*, 2009) can also occur in broccoli. In this regard, the presence of a gene encoding pheophytinase (*BoPPH*) has been demonstrated, the expression of which increases during senescence and is hormonally regulated by ethylene and cytokinins (Büchert *et al.*, 2011a; Büchert *et al.*, 2011). Also, Aiamla-or *et al.* (2012) have described an increment of PPH activity during senescence. Taken together, these results suggest the possibility that PPH would be the responsible for the dephytilation step in chlorophyll breakdown ((Aiamla-or *et al.*, 2012; Büchert *et al.*, 2011a).

The activities of the following enzymes of the catabolic pathways have not yet been measured in broccoli: pheophorbide a oxygenase and RCC reductase. However, the corresponding genes have been cloned (*BoPaO* and *BoRCCR*) and a significant increase in their expression was detected during senescence (Fukasawa *et al.*, 2010; Gómez-Lobato *et al.*, 2011). Moreover, the increment in *BoPaO* expression was delayed by cytokinins and accelerated by ethylene (Gómez-Lobato *et al.*, 2011).

Superficial color is the main quality parameter of broccoli, and one of the major technological goals is to reduce the rate of chlorophyll degradation in order to maintain green color. The two most widely utilized technologies are cooling (Pogson and Morris, 1997; Pramanik *et al.*, 2006) and controlled atmospheres (Barth *et al.*, 1993; Makhlouf *et al.*, 1990). In the first case, broccolis can maintain the color up to three weeks at 0 °C (Cho *et al.*, 2009), while the modified atmosphere packaging extends the life at 20 °C for up to a week (Eason *et al.*, 2007b; Rai *et al.*, 2009). Not only does modified atmospheres delay the breakdown of chlorophyll, but also induces the expression of several genes associated with stress (Eason *et al.*, 2007a). In the case of genes related to chlorophyll degradation, it has been shown that modified atmospheres do not affect *BoCLH1*, *BoCLH2* and *BoPPH* expression (Büchert *et al.*, 2011a), but delay the increment of *BoPaO* expression during senescence (Gómez-Lobato *et al.*, 2011).

One of the most widespread chemical treatments on postharvest technology is the use of 1-MCP, a selective blocking ethylene receptor. Several studies describe its usefulness for delaying senescence by reducing ethylene action (Cao *et al.*, 2012; Fan and Mattheis, 2000; Ma *et al.*, 2009; Watkins, 2006). These treatments can reduce degreening and chlorophyll loss during senescence of broccoli (Cefola *et al.*, 2010; Forney *et al.*, 2003; Ma *et al.*, 2010). Furthermore, treatment with 1-MCP delays the increment in the expression of *BoPaO* and *BoPPH*, and causes a lower expression of *BoRCCR* (Gómez-Lobato *et al.*, 2012). However, the same treatment does not affect the expression of *BoCLH1* and induces a higher expression of *BoCLH2,* indicating that 1-MCP selectively inhibits some but not all the genes related to chlorophyll catabolism.

Treatments with atmospheres containing ethanol at concentrations of 500 μl/L also have been effective in delaying the catabolism of chlorophyll (Fukasawa *et al.*, 2010; Han *et al.*, 2006; Xu *et al.*, 2012) and chloroplast transformation to gerontoplast (Suzuki *et al.*, 2005) during senescence. Xu *et al.* (2012) have shown that ethanol inhibits the activities of chlorophyllase, Mg-dechelatase and peroxidase, while Fukasawa *et al.* (2010) demonstrated that samples treated with ethanol have a lower expression of *BoPaO* and *BoRCCR*.

Other physical methods such as heat treatments (Lurie, 1998) were also utilized as potential postharvest methodologies in broccoli. These treatments cause a stress that modifies gene expression pattern in tissues, which in turn provokes a momentary reduction in normal metabolism (i.e. senescence) and a consequent delay in the process (Martínez and Civello,

2008). Treatments with hot air water or water can slow senescence and delay degreening up two days at 20 °C (Costa *et al.*, 2006a, Tian *et al.*, 1996). Heated heads show a lower increment of chlorophyll derivatives (Kaewsuksaeng *et al.*, 2007) and a delay in the peaks of chlorophyllase, Mg-dechelatase and peroxidase activities (Costa *et al.* 2006a; Kaewsuksaeng *et al.*, 2007). Although heat treatments reduce chlorophyllase activity, they do not have a clear and relevant effect on the expression of *BoCHL1* and *BoCHL2* related genes (Büchert *et al.*, 2011b)._On the contrary, heat has an inhibitory effect on *BoPPH* (Büchert *et al.*, 2011b) and *BoPaO* expression (Gómez-Lobato *et al.*, 2011).

It has been shown that nonlethal doses of UV-C or UV-B radiation can have a beneficial effect on the preservation of fruit and other plant products, in particular by delaying ripening and senescence (Civello *et al.*, 2007). In broccoli, a dose of 10 $KJ.m^{-2}$ of UV-C (Costa *et al.*, 2006b) or 8.8 $KJ.m^{-2}$ of UV-B (Aiamla-or *et al.*, 2010) can delay yellowing in intact heads. As in the case of heat treatments, UV radiation has a selective effect on the expression of chlorophyll degrading genes. Both UV-C and UV-B treatments do not affect *BoCHL1* and *BoCHL2* expression, but delay the increment of *BoPaO* and *BoPPH* expression during senescence (Büchert *et al.*, 2011b; Gómez-Lobato *et al.*, 2011; Aiamla-or *et al.*, 2012).

One of the determining factors of senescence is the presence of visible light, which in turn determines the level of sugars in the tissue, a factor which contributes to delaying senescence. It has been revealed that storage of broccoli in the presence of low dose of visible light delays senescence and chlorophyll degradation in approximately 2 days at 20 °C (Büchert *et al.*, 2011b). This treatment decreases *BoCHL2*, *BoPPH* (Büchert *et al.*, 2011a) and *BoPaO* (Gómez-Lobato *et al.*, 2011) expression, while not affecting the decreased *BoCHL1* expression during senescence (Büchert *et al.*, 2011a).

Postharvest life of horticultural products not only depends on treatments done after harvest but also on a range of preharvest factors, such as climate, soils, plant stress, and general crop and plant management. For example, Zaicovski *et al.* (2008) demonstrated that water stress during plant growth increases cytokinin biosynthesis and delays postharvest yellowing of broccoli florets. Plants grown at high water stress retained the green color and chlorophyll significantly better than florets obtained from plants growth at normal water regimen.

Additionally, it was described that another potentially important factor is the time of the day at which the samples are harvested (Clarkson *et al.*, 2005). In broccoli, samples harvested in the afternoon show a lower loss of color and chlorophyll degradation in comparison with those harvested in the morning (Hasperué *et al.*, 2011). The content of starch is higher in samples harvested in the late afternoon and authors hypothesize that starch degradation produces single sugars and, in this way, contribute to delay senescence (Hasperué *et al.*, 2011). What is more, most of the genes that were previously associated with chlorophyll degradation during senescence, such as *BoCLH2*, *BoPPH* and *BoPaO,* showed a lower expression or a delay in their mRNA level increments, in samples harvested at afternoon (Hasperué *et al.*, 2013).

CONCLUSION

Loss of green color during broccoli postharvest senescence involves the activation of chlorophyll degradation pathway. This metabolism is similar to that described in *Arabidopsis thaliana*, another specie of the family Cruciferae. Most of the genes identified in broccoli show considerable similarities and pattern of expression with those described in Arabidopsis. In general, the gene expression and enzymatic activities are regulated by hormones that control senescence: ethylene and cytokinins. Several types of pre and postharvest treatments may also affect the expression of these genes and, in this way, delay degradation of chlorophyll and degreening.

REFERENCES

Aiamla-or S., Nakajima T., Shigyo M., Yamauchi N. Pheophytinase activity and gene expression of chlorophyll-degrading enzymes relating to UV-B treatment in postharvest broccoli (*Brassica oleracea L. Italica Group*) florets. *Postharvest Biology and Technology*, 2012, 63, 60-66.

Aiamla-or S., Kaewsuksaeng S., Shigyo M, Yamauchi N. Impact of UV-B irradiation on chlorophyll degradation and chlorophyll-degrading enzyme activities in stored broccoli (*Brassica oleracea L. Italica Group*) florets. *Food Chemistry*, 2010, 120; 645–651.

Barth M.M., Kerbel E.L., Broussard S., Schmidt S.J. Modified Atmosphere Packaging Protects Market Quality in Broccoli Spears Under Ambient Temperature Storage. *Journal of Food Science,* 1993; 58, 1070-1072.

Benedetti C.E., Arruda P. Altering the expression of the chlorophyllase gene ATHCOR1 in transgenic Arabidopsis caused changes in the Chlorophyll-to-Chlorophyllide ratio. *Plant Physiology*, 2002, 1285, 1255-1263.

Brandis A., Vainstein A., Goldschmidt E.E. Distribution of chlorophyllase among components of chloroplast membranes in orange (*Citrus sinensis*) leaves. *Plant Physiology and Biochemitry*, 1996, 34, 49-53.

Büchert A.M., Civello P.M., Martinez G.A. Chlorophyllase versus pheophytinase as candidates for chlorophyll dephytilation during senescence of broccoli. *Journal of Plant Physiology,* 2011a, 168, 337-343

Büchert A.M., Gómez Lobato M.E., Villarreal N.M., Civello P.M., Martínez G.A. Effect of visible light treatments on postharvest senescence of broccoli (*Brassica oleracea L.*). *Journal of the Science of Food and Agriculture*, 2011b, 91, 355-361.

Cao S., Yang Z., Zheng Y. Effect of 1-methylcyclopene on senescence and quality maintenance of green bell pepper fruit during storage at 20 °C. *Postharvest Biology and Technology*, 2012, 70, 1-6.

Cefola M., Amodio M.L., Rinaldi R., Vanadia S., Colelli G. Exposure to 1-methylcyclopropene (1-MCP) delays the effects of ethylene on fresh-cut broccoli raab (*Brassica rapa L.*). *Postharvest Biology and Technology*, 2010, 58, 29-35.

Christ B., Schelbert S., Aubry S., Süssenbacher I., Müller T., Kräutler B., Hörtensteiner S. MES16, a Member of the Methylesterase Protein Family, Specifically Demethylates

Fluorescent Chlorophyll Catabolites during Chlorophyll Breakdown in Arabidopsis. *Plant Physiology,* 2011, 158, 628-641.

Chen L., Lin H, Kelkar S.M., Chang Y., Shaw J. Transgenic broccoli (*Brassica oleracea var. italica*) with antisense chlorophyllase (*BoCLH1*) delays postharvest yellowing. *Plant Science*, 2008, 174, 25-31.

Cho M.A., Hong Y.P., Choi J.W., Won Y.B., Bae D.H. Effect of packaging film and storage temperature on quality maintenance of broccoli. *Korean Journal of Horticultural Science and Technology*, 2009, 27, 128-139.

Civello P., Vicente A., Martínez G. UV-C technology to control postharvest diseases of fruits and vegetables. In "Recent Advances in Alternative Postharvest Technologies to Control Fungal Diseases in Fruits & Vegetables". 2007. Ed. Rosalba Troncoso-Rojas, Martin E. Tiznado-Hernandez y Alberto Gonzalez-Leon. *Transworld Research Network*, Kerala, India. Pages: 71-102.

Clarkson G., Rothwell S., Taylor G., End of day harvest extends shelf life. *HortScience,* 2005, 40, 1431-1435.

Costa M.L., Civello P.M., Chaves A.R., Martínez G.A., Characterization of Mg-dechelatase activity obtained from Fragaria x ananassa fruit. *Plant Physiology and Biochemistry,* 2002, 40, 111-118.

Costa M.L., Civello P.M., Chaves A.R., Martínez G.A. Effect of ethephon and 6-benzylaminopurine on chlorophyll degrading enzymes and a peroxidase-linked chlorophyll bleaching during post-harvest senescence of broccoli (*Brassica oleracea L.*) at 20 °C. *Phostharvest Biology and Thechnology*, 2004, 35, 191-199.

Costa M.L., Civello P.M., Chaves A.R., Martínez G.A. Hot air treatment decreases chlorophyll catabolism during postharvest senescence of broccoli (*Brassica oleracea L. var. italica*) heads. *Journal of the Science of Food and Agriculture*, 2006a, 86, 1125-1131.

Costa M.L., Vicente A.R., Civello P.M., Chaves A.R., Martínez G.A. UV-C treatment delays postharvest senescence in broccoli florets *Phostharvest Biology and Thechnology*, 2006b, 39, 204-210.

Deschene A., Paliyath G., Lougheed E.C., Dumbroff E.B., Thompson J.E. Membrane deterioration during postharvest senescence of broccoli florets: modulation by temperature and controlled atmosphere storage. *Postharvest Biology and Technology,* 1991, 1, 19-31.

Downs C.G., Somerfield S.D., Davey M.C. Cytokinin treatment delays senescence but not sucrose loss in harvested broccoli. *Postharvest Biology and Technology*, 1997a, 11, 93-100.

Downs C.G., Somerfield S.D., Davey M.C. Cytokinin treatment delays senescence but not sucrose loss in harvested broccoli. *Postharvest Biology and Technology*, 1997b, 11, 93-100.

Duarte-Sierra A., Corcuff R., Angers P., Arul J. Effect of heat treatment using humidified air on electrolyte leakage in broccoli florets: Temperature-Time relationships. *Acta Horticulturae*, 2012, 945,149-156.

Eason J.R., Patel D., Ryan D., Page B., Hedderley D., Watson L., West P. Controlled atmosphere treatment of broccoli after harvest delays senescence and induces the expression of novel *BoCAR* genes. *Plant Physiology and Biochemistry*, 2007a, 45, 445-456.

Eason J.R., Ryan D., Page B., Watson L., Coupe S.A. Harvested broccoli (*Brassica oleracea*) responds to high carbon dioxide and low oxygen atmosphere by inducing stress-response genes. *Postharvest Biology and Technology*, 2007b, 43, 358-365.

Engel N., Curty C., Gossauer A. Chlorophyll catabolism in Chlorella protothecoides. 8. Facts and artifacts. *Plant Physiology and Biochemistry*, 1996, 34, 77-80.

Fan X., Mattheis J.P. Reduction of ethylene-induced physiological disorders of carrots and iceberg lettuce by 1-methylcyclopropene. *HortScience*, 2000, 35, 1312-1314.

Finger F.L., Endres L., Mosquim P., Puiatti M. *Physiological changes during postharvest senescence of broccoli.* Pesquisa Agropecuária Brasileira, 1999, 34, 1565-1569.

Forney C., Song J., Fan L., Hildebrand P., Jordan M.A. Ozone and 1-methylcyclopropene alter the postharvest quality of broccoli. *journal of the American Society for Horticultural Science*, 2003, 128, 403-408.

Fukasawa A., Suzuki Y., Terai H., Yamauchi N. Effects of postharvest ethanol vapor treatment on activities and gene expression of chlorophyll catabolic enzymes in broccoli florets. *Postharvest Biology and Technology*, 2010, 55, 97-102.

Funamoto Y., Yamauchi N., Shigenaga T., Shigyo M. Effects of heat treatment on chlorophyll degrading enzymes in stored broccoli (*Brassica oleracea L.*). *Postharvest Biology and Technology,* 2002, 24, 163-170.

Funamoto Y., Yamauchi N., Shigyo M. Involvement of peroxidase in chlorophyll degradation in stored broccoli (*Brassica oleracea L.*) and inhibition of the activity by heat treatment. *Postharvest Biology and Technology*, 2003, 28, 39-46.

Funamoto Y., Yamauchi N., Shigyo M. Control of isoperoxidases involved in chlorophyll degradation of stored broccoli (*Brassica oleracea*) florets by heat treatment. *Journal of Plant Physiology*, 2006, 163 141-146.

Gómez-Lobato M.E., Civello P.M., Martinez G.A. Effects of ethylene, cytokinin, and physical treatments on BoPaO gene expression of harvested broccoli. *Journal of the Science of Food and Agriculture*, 2001, 92, 151-158.

Gómez-Lobato M.E., Hasperué J.H., Civello P.M., Chaves A.R., Martínez G.A. Effect of 1-MCP on the expression of chlorophyll degrading genes during senescence of broccoli (*Brassica oleracea L.*). *Scientia Horticulturae*, 2012, 144, 208-211.

Graham I.A., Eastmond P.J. Pathways of straight and branched chain fatty acid catabolism in higher plants. *Progress in Lipid Research*, 2002, 41, 156-181.

Halkier B.A., Gershenzon J. Biology and biochemistry of glucosinolates. *Annual Review of Plant Biology*, 2006, 57, 303-333.

Han J., Tao W., Hao H., Zhang B., Jiang W., Niu T., Li Q., Cai T. Physiology and quality responses of fresh-cut broccoli florets pretreated with ethanol vapor. *Journal of Food Science*, 2006, 71, 385-389.

Hasperué J.H., Chaves A.R., Martínez G.A. End of day harvest delays postharvest senescence of broccoli florets. *Postharvest Biology and Technology*, 2011, 59, 64-70.

Hasperué J.H., Gómez-Lobato M.E., Chaves A.R., Civello P.M., Martínez, G.A. Time of day at harvest affects the expression of chlorophyll degrading genes during postharvest storage of broccoli. *Postharvest Biology and Technology*, 2013, 82, 22-27.

Horie Y., Ito H., Kusaba M., Tanaka R., Tanaka A. Participation of chlorophyll b reductase in the initial step of the degradation of light-harvesting chlorophyll a/b-protein complexes in Arabidopsis. *Journal of Biological Chemistry*, 2009, 284, 17449-17456.

Hörstensteiner S., Wutrich K., Matile P., Ongania K., Krautler B. The key step in chlorophyll breakdown in higher plants. Cleavage of pheophorbide a macrocycle by monooxygenase. *The Journal of Biological Chemistry*, 1998, 273, 15335–15339.

Hörtensteiner S. Chlorophyll Degradation During Senescence. *Annual Review of Plant Biology*, 2006, 57, 55-57.

Hörtensteiner S. Stay-green regulates chlorophyll and chlorophyll-binding protein degradation during senescence. *Trends in Plant Science*, 2009, 14, 155-162.

Hörtensteiner S., Kräutler B. Chlorophyll breakdown in higher plants. *Biochimica et Biophysica Acta- Bioenergetics*, 2011, 1807, 977-988.

Hörtensteiner S., Schelbert S., Aubry S., Burla B., Agne B., Kessler F., Krupinska K. Pheophytin pheophorbide hydrolase (Pheophytinase) is involved in chlorophyll breakdown during leaf senescence in Arabidopsis. *The Plant Cell*, 2009, 21, 767-785.

Iturraspe J., Engel N., Gossauer A. Chlorophyll catabolism. Isolation and structure elucidation of chlorophyll b catabolites in *Chlorella protothecoides*. *Phytochemistry*; 1994, 35, 1387-1390.

Jeffery E., Araya M. Physiological effects of broccoli consumption. *Phytochemistry Reviews,* 2009, 8, 283-298.

Kaewsuksaeng S., Yamauchi N., Funamoto Y., Aiamla-Or S., Shigyo M., Kanlayanarat S. Partially purification of Mg-dechelatase in relation to chlorophyll degradation in broccoli (*Brassica oleracea L. Italica Group*) florets. *Acta Horticulturae*, 2010, 875, 509-514.

Kaewsuksaeng S., Yamauchi N., Funamoto Y., Shigyo M., Kanlavanarat S. Effect of Mg-dechelation activity on chlorophyll degradation in stored broccoli florets. *Acta Horticulturae,* 2006, 746, 705-709.

King G.A., Morris S.C. Physiological changes of broccoli during early postharvest senescence and through the preharvest-postharvest continuum. *journal of the American Society for Horticultural Science,* 1994, 119, 270-275.

Ku V.V.V., Wills R.B.H. Effect of 1-methylcyclopropene on the storage life of broccoli. *Postharvest Biology and Technology*, 1999, 17, 127-132.

Kunieda T., Amano T., Shioi Y. Search for chlorophyll degradation enzyme, Mg-dechelatase, from extracts of Chenopodium album with native and artificial substrates. *Plant Science,* 2005, 169, 177-183.

Lee G., Chepyshko H., Chen H., Chu C., Chou Y, Akoh C., Shaw J. Genes and biochemical characterization of three novel chlorophyllase isozymes from *Brassica oleracea. Journal of Agricultural and Food Chemistry*, 2010, 58, 8651-8657.

Lemoine L., Civello P., Chaves A., Martínez G. Hot air treatment delays senescence and maintains quality of fresh-cut broccoli florets during refrigerated storage. *Food Science and Technology*;2009; 42, 1076-1081.

Lurie S. Postharvest heat treatments. *Postharvest Biology and Technology*, 1998, 14, 257-269.

Ma G., Wang R., Wang C, Kato M., Yamawaki K., Qin F, Xu, H. Effect of 1-methylcyclopropene on expression of genes for ethylene biosynthesis enzymes and ethylene receptors in post-harvest broccoli. *Plant Growth Regulation*, 2009, 57, 223-232.

Ma G., Zhang L., Kato M., Yamawaki K., Asai T., Nishikawa F., Ikoma Y., Matsumoto H. Effect of 1-methylcyclopropene on the expression of genes for ascorbate metabolism in postharvest broccoli. *Postharvest Biology and Technology*, 2010, 58, 121-128.

Maeda Y., Kurata H., Adachi M., Shimokawa K. Chlorophyll catabolism in ethylene-treated *Citrus unshiu* fruits. *Journal of Japanese Society for Horticulture Science*, 1998, 67, 497-502.

Makhlouf J., Willemot C., Couture R., Arul J., Castaigne F. Effect of low temperature and controlled atmosphere storage on the membrane lipid composition of broccoli flower buds. Scientia Horticulturae, 1990, 42, 9-19.

Martínez G., Civello P., Chaves A., Añón, M. Characterization of peroxidase-mediated chlorophyll bleaching in strawberry fruit. *Phytochemistry*, 2001, 58, 379-387.

Martínez G, Civello P. Effect of heat treatments on gene expression and enzyme activities associated to cell wall degradation in strawberry fruit. *Postharvest Biology and Technology*, 2008, 49, 38-45.

Matile P., Ginsburg S., Schellenberg M., Thomas H. Catabolites of chlorophyll in senescent leaves. *Journal of Plant Physiology*, 1987, 129, 219.

Matile P., Hörtensteiner S., Thomas H. Chlorophyll degradation. *Annual Review of Plant Physiology and Plant Molecular Biology,* 1999, 50, 67-95.

Meguro M., Ito H., Takabayashi A., Tanaka R., Tanaka A. Identification of the 7-hydroxymethyl chlorophyll a reductase of the chlorophyll cycle in Arabidopsis. *The Plant Cell,* 2011, 23, 3442-3453.

Ni X., Quisenberry S., Markwell J., Heng-Moss T., Higley L., Baxendale F., Sarath G., Klucas R. In vitro enzymatic chlorophyll catabolism in wheat elicited by cereal aphid feeding. *Entomologia Experimentalis et Applicata*, 2001, 101, 159-166.

Nishikawa F., Iwama T., Kato M., Hyodo H., Ikoma Y., Yano M. Effect of sugars on ethylene synthesis and responsiveness in harvested broccoli florets. *Postharvest Biology and Tecnology*, 2005, 36, 157–165.

Page T., Griffiths G., Buchanan-Wollaston V. Molecular and biochemical characterization of postharvest senescence in Broccoli. *Plant Physiology*; 2001, 125, 718-727.

Pogson B.J., Morris S. Consequences of cold storage of broccoli on physiological and biochemical changes and subsequent senescence at 20 ºC. *Journal of the American Society for Horticultural Science*, 1997, 122, 553-558.

Pogson B., Downs C., Davies K. Differential expression of two 1-aminocyclopropane-1-carboxylic acid oxidase genes in broccoli after harvest. *Plant Physiology*, 1995, 108, 651-657.

Pramanik B.K., Matsui T., Suzuki H., Kosugi Y. Compositional and some enzymatic changes relating to sugar metabolism in broccoli during storage at 1°C and subsequent senescence at 20°C. *Acta Horticulturae*, 2006, 706, 219-227.

Pruzinskà A., Tanner G., Anders I., Roca M., Hörtensteiner S. Chlorophyll breakdown: pheophorbide a oxygenase is a rieske-type ironsulfur protein, encoded by the accelerated cell death 1 gene. *Proceedings of the National Academy of Sciences of the United States of America,* 2003, 100, 15259-15264.

Pruzinská A., Tanner G., Aubry S., Anders I., Moser S., Muller T., Ongania K., Krautler B., Youn J., Liljegren S., Hörtensteiner S. Chlorophyll breakdown in senescent Arabidopsis leaves. characterization of chlorophyll catabolites and of chlorophyll catabolic enzymes involved in the degreening reaction. *Plant Physiology*, 2005, 139, 52-63.

Rai D.R., Jha S.N., Wanjari O.D., Patil R.T. Chromatic changes in broccoli (*Brassica oleracea italica*) under modified atmospheres in perforated film packages. *Food Science and Technology International*, 2009, 15, 387-395.

Roca M., James C., Pružinská A., Hörtensteiner S., Thomas H., Ougham H. Analysis of the chlorophyll catabolism pathway in leaves of an introgression senescence mutant of *Lolium temulentum. Phytochemistry*, 2004, 65, 1231-1238.

Rodoni S., Mhlecker W., Anderl M., Krutler B., Moser D. Chlorophyll breakdown in senescent chloroplasts. Cleavage of pheophorbide a in two enzymic steps. *Plant Physiology,* 1997, 115, 669.

Schelbert S., Aubry S., Burla B., Agne B., Kessler F., Krupinska K., Hörtensteiner S. Pheophytin pheophorbide hydrolase (pheophytinase) is involved in chlorophyll breakdown during leaf senescence in Arabidopsis. *Plant Cell,* 2009, 21, 767-785.

Schenk N., Schelbert S., Kanwischer M., Goldschmidt E., Dörmann P., Hörtensteiner S. The chlorophyllases *AtCLH1* and *AtCLH2* are not essential for senescence-related chlorophyll breakdown in *Arabidopsis thaliana. FEBS Letters*, 2007, 581, 5517-5525.

Shioi Y., Watanabe K., Takamiya K. Enzymatic conversion of pheophorbide a to the precursor of pyropheophorbide a in leaves of *Chenopodium album. Plant Cell Physiology,* 1996, 37, 1143.

Singh J., Upadhyay A.K., Prasad K., Bahadur A., Rai M. Variability of carotenes, vitamin C, E and phenolics in *Brassica* vegetables. *Journal of Food Composition and Analysis,* 2007, 20, 106-112.

Suzuki T., Kunieda T., Murai F., Morioka S., Shioi Y. Mg-dechelation activity in radish cotyledons with artificial and native substrates, Mg-chlorophyllin a and chlorophyllide a. *Plant Physiology and Biochemistry*, 2005, 43, 459-464.

Suzuki T., Shioi Y. Re-examination of Mg-dechelation reaction in the degradation of chlorophylls using chlorophyllin a as a substrate. *Photosynthesis Research*, 2002, 74, 217-223.

Takahashi Y., Adachi M., Furuta M., Yamamoto H., Kurata H., Azuma R., Miyoshi H., Shimokawa K. Chlorophyll Catabolism in Broccoli (*Brassica oleracea L.*) Flower Buds : An Involvement of Mg-dechelatase. *Journal of the Japanese Society for Horticultural Science,* 2001, 70, 121-125.

Takamiya K., Tsuchiya T., Ohta H. Degradation pathway(s) of chlorophyll: what has gene cloning revealed? *Trends in Plant Science*, 2000, 5, 426-431.

Thompson J.E., Froese C.D., Madey E., Smith M.D., Hong Y. Lipid metabolism during plant senescence. *Progress in Lipid Research*, 1998, 37, 119-141.

Tian M., Downs C., Lill R., King G. A role for ethylene in the yellowing of broccoli after harvest. *Journal of the American Society for Horticultural Science,* 1994, 119, 276-281.

Tian M., Woolf A., Bowen J., Ferguson I. Changes in colour and chlorophyll fluorescence of broccoli florets following hot water treatment. *Journal of American Socity for Horticulture Science*, 1996, 121, 310–313.

Tsuchiya T., Ohta H., Okawa K., Iwamatsu A., Shimada H., Masuda T., Takamiya K . Cloning of chlorophyllase, the key enzyme in chlorophyll degradation: Finding of a lipase motif and the induction by methyl jasmonate. *Proceedings of the National Academy of Sciences of the United States of America,* 1999, 96, 15362-15367.

Vicentini F., Hörtensteiner S., Schellenberg M., Thomas H., Matile P. Chlorophyll breakdown in senescent leaves: identification of the lesion in a stay-green genotype of *Festuca pratensis. New Phytologist*, 1995, 129, 247-252.

Wang Y.T., Yang C.Y., Chen Y., Lin Y., Shaw, J. Characterization of senescence-associated proteases in postharvest broccoli florets. *Plant Physiology and Biochemistry,* 2004, 42, 663-670.

Watkins C.B. The use of 1-methylcyclopropene (1-MCP) on fruits and vegetables. *Biotechnology Advances*, 2006, 24, 389-409.

Wold A.B., Lea P., Jeksrud W.K., Hansen M., Rosenfeld H.J., Baugerød H., Haffner K. Antioxidant activity in broccoli cultivars (*Brassica oleracea var. italica*) as affected by storage conditions. *Acta Horticulturae,* 2006, 706, 211-217.

Wüthrich K.L., Bovet L., Hunziker P.E., Donninson I.S., Hörtensteiner S. Molecular cloning, functional expression and characterisation of RCC reductase involved in chlorophyll catabolism. *The Plant Journal*, 2000, 21, 189-198.

Xu F., Chen X., Jin P., Wang X., Wang J., Zheng Y. Effect of ethanol treatment on quality and antioxidant activity in postharvest broccoli florets. *European Food Research and Technology*, 2012, 5, 793-800.

Zaicovski C.B., Zimmerman T., Nora L., Nora F.R., Silva J.A., Rombaldi C.V. Water stress increases cytokinin biosynthesis and delays postharvest yellowing of broccoli florets. *Postharvest Biology and Technology*, 2008, 49, 436-439.

Zapata P.J., Tucker G.A., Valero D., Serrano M. Quality parameters and antioxidant properties in organic and conventionally grown broccoli after pre-storage hot water treatment. *Journal of the Science of Food and Agriculture*. In Press.

Zhuang H., Hildebrand D.F., Barth M.M. Temperature influenced lipid peroxidation and deterioration in broccoli buds during postharvest storage. *Postharvest Biology and Technology*, 1997, 10, 49-58.

In: Brassicaceae
Editor: Minglin Lang
ISBN: 978-1-62808-856-4

Chapter 5

MINI-REVIEW OF THE MOLECULAR PROPERTIES AND PHYSIOLOGICAL FUNCTIONS OF NON-PHOTOCONVERTIBLE WATER-SOLUBLE CHLOROPHYLL-BINDING PROTEINS (WSCPS) IN BRASSICACEAE PLANTS

Shigekazu Takahashi and Hiroyuki Satoh*
Department of Biomolecular Science, Toho University, Chiba, Japan

ABSTRACT

This mini-review provides updated information about the molecular properties and the deduced physiological functions of water-soluble chlorophyll-binding proteins (WSCPs). Chlorophylls (Chls) are essential pigments in photosynthesis, and all Chl-binding proteins functioning in photosynthesis are hydrophobic proteins located in thylakoids. Unique hydrophilic non-photoconvertible Chl-binding proteins called Class II WSCPs have been extracted from various Brassicaceae plants. In contrast to photosynthetic Chl-binding proteins, WSCPs contain only Chls, and no carotenoids. Carotenoids are able to moderate the generation of reactive oxygen species (ROS) produced during photoreaction, but only WSCPs are able to suppress the generation of ROS derived from excited Chls under light conditions. In plant cells, WSCPs are located in the endoplasmic reticulum body, a unique organelle found only in Brassicaceae plants and thought to participate in defense against pests and/or pathogen attacks. These observations indicate that WSCPs may function as Chl scavengers when plant cells are injured. All of the WSCPs cloned from Brassicaceae plants to date possess a signature motif of Kunitz-type trypsin inhibitor (KTI). Another important aspect of WSCPs is their protease inhibitory activity. A recent proteomic profiling study demonstrated that a trypsin inhibitor activity of BnD22 (a Brassica napus WSCP) was important to protect younger tissues under nitrogen starvation. Moreover, WSCPs are stress-inducible

---

* Corresponding author: Hiroyuki Satoh, Department of Biomolecular Science, Faculty of Science, Toho University, 2-2-1 Miyama, Funabashi, Chiba 274-8510, Japan; Tel.: +81-47-472-7532; Fax: +81-47-472-7532; E-mail: hsatoh@biomol.sci.toho-u.ac.jp

proteins, and thus it was hypothesized that WSCPs may be bifunctional molecules, with roles in both stress-induced Chl scavenging and protease inhibition.

Keywords: Water-soluble chlorophyll-binding protein, WSCP, Class II WSCP, Kunitz-type trypsin inhibitor, ER body, chlorophyll scavenger

INTRODUCTION

Photosynthesis is an important biological process whereby the energy of sunlight is harvested and converted into storable chemical compounds such as carbohydrates and lipids. It is not an overstatement to say that photosynthesis provides almost all of the energy required for life on Earth. There are several types of photosynthesis systems, with the best known and studied being oxygenic photosynthesis. All oxygenic phototrophic organisms, including cyanobacteria, algae, moss, ferns, and plants, possess chlorophylls (Chls) to absorb light energy. Various Chls have also been found in photoautotrophic organisms, but land plants contain only two types of Chls, Chl *a* and Chl *b*.

Because Chl and its metabolic intermediates, such as protoporphyrinogen IX and pheophorbide *a*, are phototoxic, a non-enzymatic excess accumulation of these pigments generates huge amounts of reactive oxygen species (ROS), which causes cell death. The Chl metabolism in plant cells must therefore be strictly regulated. Almost all of the major synthetic and degradation pathways of Chls have been elucidated (for review see Tanaka and Tanaka, 2007; Hörtensteiner and Kräutler, 2011; Hörtensteiner, 2012). However, knowledge of the Chl-degradation pathway in injured cells is still limited.

All Chl-binding proteins involved in photosynthesis are hydrophobic membranous proteins, and all of them function in thylakoids. In contrast to these Chl-binding proteins, highly hydrophilic Chl-binding proteins known as water-soluble chlorophyll-binding proteins (WSCPs) have been extracted from several land plants classified into the families Brassicaceae, Chenopodiaceae, Amaranthaceae and Polygonaceae (for review see Satoh *et al.*, 2001). The first identified WSCP from *Chenopodium album* exhibited a unique photoconvertibility (Yakushiji *et al.*, 1963). Photoconvertible WSCPs were later found in Amaranthaceae and Polygonaceae plants. Non-photoconvertible WSCPs, on the other hand, have been extracted only from Brassicaceae plants.

The WSCPs are categorized into two classes based on their photoconvertibility: Chenopodiaceae-type (Class I) and Brassicaceae-type (Class II) (for review see Satoh *et al.*, 2001). The amino acid sequences of Brassicaceae-WSCPs differ from those of Chenopodiaceae-WSCPs, implying that the biological function(s) of the two types of WSCPs may be different. At the least, these WSCPs possess different evolutional histories. Although the biological function(s) of these WSCPs have been an enigma, several interesting biological functions of Brassicaceae WSCPs have been proposed in the last decade. This mini-review describes the current knowledge of the biochemical properties and proposed biological functions of the Brassicaceae WSCPs (hereafter referred to as simply WSCPs).

WSCPs in Brassicaceae Plants

In 1971, Murata *et al.* extracted a WSCP from *Brassica oleracea* var. *botrys* (cauliflower). WSCPs were subsequently isolated from several *Brassica* plants; e.g., *Brassica nigra* (black mustard) (Murata and Murata, 1971), *Brassica napus* (rapeseed) (Reviron *et al.*, 1992), *Brassica oleracea* var. *gemmifera* (Brussels sprouts) (Kamimura *et al.*, 1997), and *Brassica oleracea* var. *acephala* (kale) (Horigome *et al.*, 2003). Based on the absorption peak at the red band, these *Brassica*-WSCPs were called CP674 or CP673; however, this nomenclature could not distinguish individual WSCPs. To avoid misunderstanding, we describe these WSCPs as cauliflower WSCP (cauWSCP), black mustard WSCP (bmuWSCP), Brussels sprouts WSCP (BoWSCP) and kale WSCP (kalWSCP) in this mini-review. Among *Brassica* species, it is interesting that WSCPs have been found only in the species possessing a B and/or C genome, and not in those with an A genome (Takamiya and Yakushiji, 1976).

In addition to *Brassica*-WSCPs, other Brassicaceae-type WSCPs, including *Lepidium*-WSCPs (LvWSCPs, which were formerly called CP661 [Murata and Murata, 1971; Murata and Ishikawa, 1981] and CP663 [Murata and Ishikawa, 1981; Itoh *et al.*, 1982]), *Raphanus*-WSCPs (RsWSCPs) (Shinashi *et al.*, 2000; Takahashi *et al.*, 2012a) and the *Arabidopsis*-WSCP (AtWSCP) (Bektas *et al.*, 2012), have been isolated from *Lepidium virginicum* (Virginia pepperweed), *Raphanus sativus* (Japanese radish and Japanese wild radish) and *Arabidopsis thaliana* (mouse-ear cress).

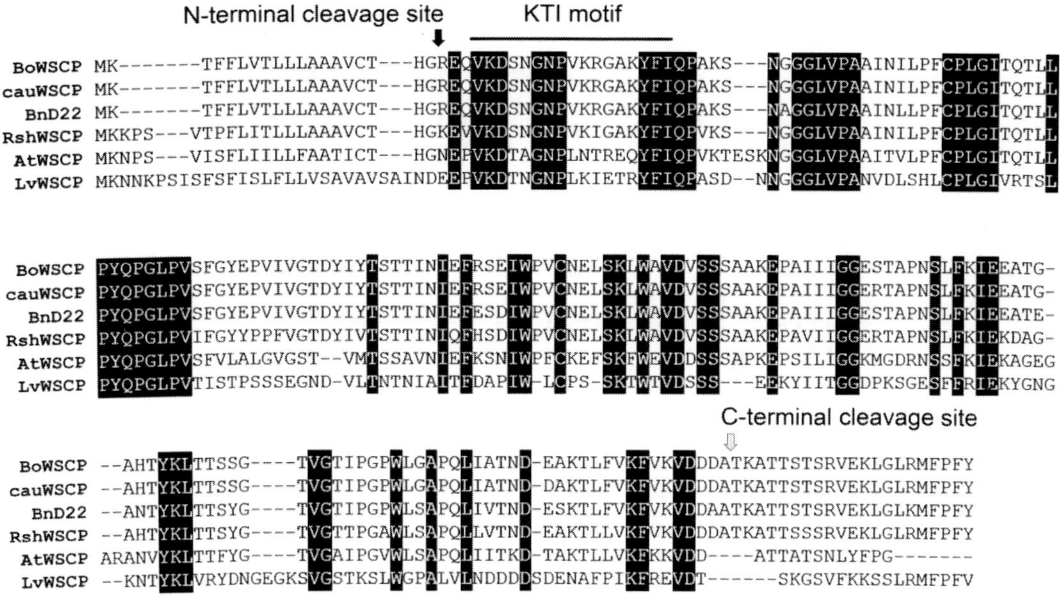

Figure 1. Alignment of deduced amino acid sequences of WSCPs. The amino acid sequences of WSCPs from rapeseed (BnD22), cauliflower (cauWSCP), *Arabidopsis thaliana* (AtWSCP), Brussels sprouts (BoWSCP), Japanese wild radish (RshWSCP) and Virginia pepperweed (LvWSCP) were aligned using the Clustal X program. The amino acid residues conserved among all WSCPs are highlighted with a black background. The black and gray arrows indicate the N- and C-terminal cleavage sites, respectively. The Kunitz trypsin inhibitor (KTI) motif is indicated by a black bar.

Several genes and cDNAs encoding WSCPs (BnD22, Downing *et al.*, 1992; cauWSCP, Satoh *et al.*, 1998; AtWSCP, Bektas *et al.*, 2012; BoWSCP, Takahashi *et al.*, 2012b; RshWSCP, Takahashi *et al.*, 2012a; and LvWSCP, our unpublished data) have been cloned thus far. Figure 1 shows the sequence alignment of deduced amino acid residues of the WSCPs.

BASIC MOLECULAR PROPERTIES OF WSCPS

All native WSCPs (i.e., those with the Chl-binding form) characterized thus far are homo-tetramers, which are composed of 19-20 kDa subunits and a few Chls (Kamimura *et al.*, 1997; Nishio and Satoh, 1997; Shinashi *et al.*, 2000; Takahashi *et al.*, 2012b). The Chl compositions of the WSCPs are diverse. The Chl *a*/*b* ratios of cauWSCP, BnWSCP, and kalWSCP were estimated to be around 6.0 (Murata *et al.*, 1971; Murata and Murata, 1971; Horigome *et al.*, 2003), and those of BoWSCP and RshWSCP were estimated to be over 10.0 (Kamimura *et al.*, 1997; Takahashi *et al.*, 2012b).

In contrast, the Chl *a*/*b* ratio of LvWSCP has been reported to be 1.0-3.5 in *Lepidium virginicum* (Murata and Murata, 1971; Murata and Ishikawa, 1981; Itoh *et al.*, 1982). Murata and Ishikawa (1981) reported that the Chl *a*/*b* ratio of a LvWSCP purified from Virginia pepperweed collected in California in the US was 1.0 and that of Virginia pepperweed harvested in Chiba, Japan was 1.6-1.9. Moreover, Itoh *et al.* found that the Chl *a*/*b* ratio of an LvWSCP extracted from leaves was 1.5-1.7 and that from stems was 3.4-3.5 (Itoh *et al.*, 1982). It is unclear whether these LvWSCPs are isozymes or isoforms.

Based on the Chl *a*/*b* ratio, WSCPs can be categorized into two subclasses: Class IIA (Chl *a*/*b* = 6.0 ~ 10.0) and Class IIB (Chl *a*/*b* 1.0 ~ 3.5) (for review see Satoh *et al.*, 2001). All of the known *Brassica*-WSCPs and *Raphanus*-WSCPs are categorized into Class IIA WSCPs, while the LvWSCPs are the only Class IIB WSCPs thus far. It should be noted that the Chl *a*/*b* ratios of AtWSCP have not yet been characterized and thus the subclass of the AtWSCP are unknown at this time. The above information is summarized in Table 1.

Table 1. The Brassicaceae (Class II) WSCPs

WSCP Name	Alias name	Species name	Popular name	Subclass
CauWSCP	CP673	*Brassica oleracea* var. *botrys*	Caulifrower	A
BoWSCP	CP674	*Brassica oleracea* var. *gemmifera*	Brussels sprouts	A
KalWSCP	CP673	*Brassica oleracea* var. *acephala*	Kale	A
BmuWSCP	CP673	*Brassica nigra*	Black mustard	A
BnD22		*Brassica napus*	Rapeseed	A
RsdWSCP		*Raphanus sativus* var. *hortensis*	Japanese radish	A
RshWSCP		*Raphanus sativus* var. *raphanistroides*	Japanese wilde radish	A
LvWSCP	CP661 CP663	*Lepidium virginicum*	Virginia pepperweed	B
AtWSCP		*Arabidopsis thaliana*	Mouse-ear cress	Unknown

The Apo-form of recombinant WSCPs expressed in *E. coli* (e.g., cauWSCP, Satoh *et al.*, 1998; AtWSCP, Bektas *et al.*, 2012; BoWSCP, Takahashi *et al.*, 2012b; and LvWSCP, our unpublished data) could be reconstituted into the holo-form (i.e., a homo-tetramer) *in vitro* by mixing purified Chls in a buffer containing 20%-40% organic solvent such as methanol or ethanol. The recombinant cauWSCP could be reconstituted to the holo-form by means of not only Chl but also bacteriochlorophyll *a*, a Zn derivative of Chl *a* (Schmidt *et al.*, 2003). However, the recombinant protein failed to bind pheophytin *a*, which is a porphyrin lacking a central coordinated metal, indicating that the existence of a central metal ion of Chls or its derivatives is essential for binding (Schmidt *et al.*, 2003).

In addition, the recombinant cauWSCP also binds several Chl precursors lacking a phytol chain (i.e., chlorophyllide [Chlide] *a*, Chlide *b* and Mg-protoporphyrin IX); however, none of the cauWSCPs with the pigment formed a tetramer, indicating that the oligomerization of WSCPs requires pigments possessing a phytol chain (Schmidt *et al.*, 2003).

We show three typical absorption spectra and peak wavelengths of WSCPs (BoWSCP, RsdWSCP and LvWSCP) in Figure 2 and Table 2, respectively. The absorption spectra and peaks of WSCPs reflect the Chl content and Chl *a*/*b* ratio of each WSCP. The absorption peak at 470 nm found in LvWSCP is derived from Chl *b*, and the other peaks (excluding the peak at 699 nm in RsdWSCP) are derived from Chl *a*. To the best of our knowledge, there has been no report describing a pigment responsible for a peak at 669 nm; this peak is found only in RsdWSCP.

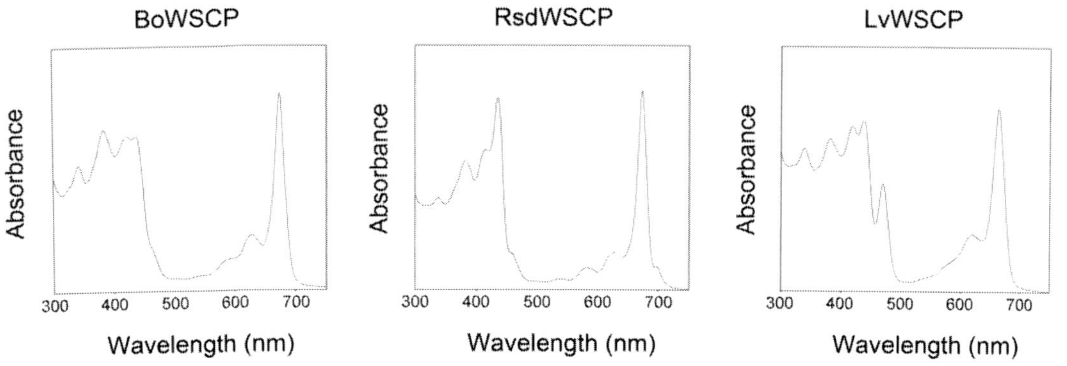

Figure 2. Absorption spectra (300-750 nm) of BoWSCP, RsdWSCP and LvWSCP. The UV-visible absorption spectra of these WSCPs were measured in 20 mM Tris-HCl (pH 8.0) at 25°C.

Table 2. Absorption peaks (nm) of WSCPs in 20 mM Tris-HCl (pH 8.0) at 25°C

	Soret					*Q* band		
BoWSCP	342	383	422	437		629	673	
RsdWSCP	338	382	416	436		628	673	699
LvWSCP	339	382	419	438	470	618	663	

CHL-BINDING AND SUPPRESSION OF THE ROS GENERATION MECHANISMS OF WSCPS

It is well known that all Chl-binding proteins (e.g., light-harvesting complexes) functioning in photosynthesis contain not only Chls but also carotenoids, which are important pigments for the heat dissipation of excess light energy captured by Chls. Although WSCPs bind only Chls, WSCPs are able to suppress the generation of ROS derived from excited Chls under light conditions (Schmidt *et al.*, 2003).

Recently, Horigome *et al.* (2007) revealed the molecular structure of an LvWSCP by conducting an X-ray crystal structure analysis and explained how WSCPs suppress the generation of ROS. Briefly, a holo-form of LvWSCP forms one hydrophobic cavity the size of which corresponds to four Chls at the interior portion of the complex, and the holo-form of LvWSCP used this cavity to capture four Chls. Although the cavity mentioned above possesses four pore-like vents, the vents are too small to contact outside of the cavity, and thus interaction between Chls captured in this cavity and molecules located outside of the cavity cannot occur. Therefore, excited Chls in the LvWSCP complex cannot transfer their own energy to a molecular oxygen located outside the complex, and thus a generation of ROS is suppressed.

WSCPs May be Chl Scavengers

The existence of a Chl carrier protein which is able to interact with chlorophyllase, an enzyme functioning in the first step of a Chl-degradation pathway (Matile *et al.*, 1997), has long been predicted. Since WSCPs are hydrophilic Chl-binding proteins, it has been hypothesized that WSCPs may be Chl carriers (Kamimura *et al.*, 1997). Chls and their catabolizing enzymes in the Chl-degradation pathway are located in chloroplasts (for review see Hörtensteiner, 2012). If WSCPs act as Chl carriers in the pathway, WSCPs must be located in chloroplasts. However, none of the WSCPs characterized thus far possess a transit peptide into the chloroplast; rather, they possess a deduced signal peptide targeting the endoplasmic reticulum (ER) (for review see Satoh *et al.*, 2001; Takahashi *et al.*, 2012a, b). In addition, Sakuraba *et al.* recently demonstrated that the chlorophyll breakdown in senescent leaves occurred within a thylakoidal protein complex composed of six proteins (i.e., NONYELLOW COLORING1 (NYC1), NYC1-LIKE (NOL), pheophytinase (PPH), pheophorbide a oxygenase (PAO), red chlorophyll catabolite reductase (RCCR) and STAY-GREEN (SGR)) without WSCP. Moreover, Damaraju *et al.* (2011) reported that cauWSCP overexpression in tobacco does not affect Chl metabolism during the life cycle, including in senescent leaves. These recent observations imply that WSCPs do not function as a Chl carrier in the Chl-catabolic pathway under senescent conditions.

Through the analysis of a transgenic *Arabidopsis thaliana*, which expresses a fluorescent protein fused with a conserved signal peptide of *Brassica*-WSCPs, we demonstrated that the signal peptide targets the ER body, which is one of the derivatives of ER (Takahashi *et al.*, 2012b). It should be noted that WSCPs also possess a C-terminal extension region (Ilami *et al.*, 1997; Takahashi *et al.*, 2012b), which is removed in the mature form. Ilami *et al.* (1997) hypothesized that this region may be a trailer peptide that targets chloroplasts via the ER. We examined whether the C-terminus extension is a trailer peptide by a transgenic study, and found that the C-terminus extension was not the trailer peptide targeting plastids (our unpublished data). We thus concluded that WSCPs are located in the ER body.

It is interesting that the ER body has been found only in Brassicaceae plants (for review see Yamada *et al.*, 2011). In comparison to other organelles, such as the chloroplast and mitochondria, the ER body is a plastic organelle (i.e., the ER body easily changes its own formation and numbers). Although the number of ER bodies in the leaves declines with the progression of growth under non-stress conditions, formation of the ER body is induced by wounding and treatment with methyl jasmonate, which is a hormone that mediates the resistance response to pests and pathogen attacks (for review see Yamada *et al.*, 2011). Like the ER body, WSCPs are also markedly induced by methyl jasmonate treatment (Desclos *et al.*, 2008). Moreover, various environmental stresses including drought (Downing *et al.*, 1992; Reviron *et al.*, 1992), salinity (Reviron *et al.*, 1992), heat (Annamalai and Yanagihara, 1999), and nutrient deficiency (Desclos *et al.*, 2008) also enhance the expression and accumulation of WSCPs.

In *Arabidopsis thaliana*, one of the major contents of the ER body has been reported to be PYK10, which is one of the β-glucosidases and is considered to function in the defense response against pest/pathogen attack by producing toxic products (for review see Yamada *et al.*, 2011). It is interesting that substrates and enhancers of PYK10 are located in the vacuole and cytosol, respectively. Yamada *et al.* (2011) thus proposed that the ER body may be a unique defense system established by Brassicaceae plants. According to the proposed function of the ER body and molecular properties of WSCPs described above, we proposed that WSCPs may be Chl scavengers that function to suppress the oxidative stress induced by Chls when the plant cells are injured (Takahashi *et al.*, 2012b) (Figure 3).

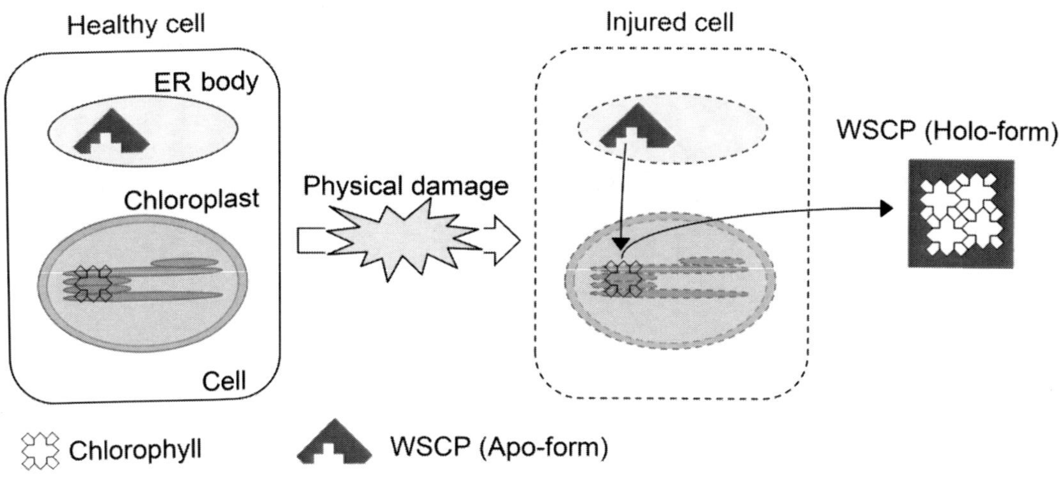

Figure 3. Dynamic model of the Chl-binding mechanism of WSCPs in a plant cell. In a healthy cell, apo-WSCPs are located in the ER body. When cells are injured, apo-WSCPs are released from the ER body and then immediately scavenge Chls located in the damaged thylakoid to suppress the production of reactive oxygen species (ROS) derived from the Chls.

OTHER ASPECTS OF WSCPS

All of the amino acid sequences of the WSCPs characterized thus far possess the signature motif of a Kunitz-type trypsin inhibitor (KTI) on an N-terminus region (Figure 1). In fact, cauWSCP was cloned as a homologous protein of BnD22 (a drought- and salinity-stress-induced 22-kDa protein of rapeseed [*Brassica napus*]) (Nishio and Satoh, 1997; Satoh *et al.*, 1998), which is one of the KTI family proteins. In addition, recombinant BnD22 could bind Chls (Satoh *et al.*, 1998), confirming that the BnD22 is a WSCP of rapeseed. Ilami *et al.* (1997) proposed that BnD22 may function in the suppression of a proteinase activity in drought-adapted leaves and delay the leaf senescence. Nonetheless, these initially characterized WSCPs possess a KTI motif, and it has been reported that BnD22 and native cauWSCP did not inhibit or only slightly inhibited the proteinase activity of trypsin and chymotrypsin *in vitro* (Ilami *et al.*, 1997; Nishio and Satoh, 1997).

However, Desclos *et al.* (2008) demonstrated that BnD22 exhibited trypsin inhibitor activity *in vivo*. They also discovered that the expression levels of BnD22 and various WSCP homologous proteins were induced by nitrogen starvation, especially in young leaves, and thus they suggested that BnD22 possesses dual functions (Chl scavenging and trypsin inhibition) to protect younger tissues by maintaining metabolism in *Brassica napus*. In addition, Halls *et al.* (2006) found that the AtWSCP precursor could inhibit papain, which is one of the cysteine proteases. Bektas *et al.* (2012) found that AtWSCP is located in the transmitting tract in the gynoecium and silique, and they thus proposed that AtWSCP may function not only as a Chl carrier but also as a protease inhibitor. We expect that the proteinase inhibitory activity of WSCPs will be revealed in future studies.

In conclusion, WSCPs exhibit many interesting properties, functioning as stress-induced, Chl-binding, ROS-suppressing, and proteinase-inhibitory proteins. Non-photoconvertible WSCPs have been found thus far only in Brassicaceae plants, and these non-photoconvertible WSCPs are located in the ER body, which is found only in Brassicaceae plants. Further research on WSCPs is expected to open a new field focusing on how Brassicaceae plants adapt to their environmental stresses.

REFERENCES

Annamalai, P., Yanagihara, S., (1999) Identification and characterization of a heat-stress incuced gene in cabbage encodes a Kunitz type protease inhibitor. *J. Plant Physiol.* 155, 226-233.

Bektas, I., Fellenferg, C., Paulsen, H., (2012) Water-soluble chlorophyll protein (WSCP) of Arabidopsis is expressed in the gynoecium and developing silique. *Planta* 236, 251-259.

Damaraju, S., Schlede, S., Eckhardt, U., Lokstein, H., Grimm, B., (2011) Functions of the water soluble chlorophyll-binding protein in plants. *J. Plant Physiol.* 168, 1444-1451.

Desclos, M., Dubousset, L., Etienne, P., Le Caherec, F., Satoh, H., Bonnefoy, J., Ourry, A., Avice, JC., (2008) A proteomic profiling approach to reveal a novel role of *Brassica napus* drought 22 kD/water-soluble chlorophyll-binding protein in young leaves during nitrogen remobilization induced by stressful conditions. *Plant Physiol.* 147, 1830-1844.

Downing, WL., Mauxion, F., Fauvarque, MO., Reviron, MP., de Vienne, D., Vartanian, N., Giraudat, J., (1992) A *Brassica napus* transcript encoding a protein related to the Künitz protease inhibitor family accumulates upon water stress in leaves, not in seeds. *Plant J.* 2, 685-693.

Halls, CE., Rogers, SW., Oufattole, M., Ostergard, O., Sevensson, B., Rogers, JC., (2006) A Kunitz-type cysteine protease inhibitor from cauliflower and Arabidopsis. *Plant Sci* 170, 1102-1110.

Horigome, D., Satoh, H., Uchida, A., (2003) Purification, crystallization and preliminary X-ray analysis of a water-soluble chlorophyll protein from *Brassica oleracea* L. var. *acephala* (kale). *Acta Crystallogr. D Biol. Csystallogr.* 59, 2283-2285.

Horigome, D., Satoh, H., Itoh, N., Mitsunaga, K., Oonishi, I., Nakagawa, A., Uchida, A., (2007) Structural mechanism and photoprotective function of water-soluble chlorophyll-binding protein. *J. Biol. Chem.* 282, 6525-6531.

Hörtensteiner, S., Kräutler, B., (2011) Chlorophyll breakdown in higher plants. *Biochim. Biophys. Acta* 1807, 977-988.

Hörtensteiner, S., (2012) Update on the biochemistry of chlorophyll breakdown. *Plant Mol. Biol.* Jul. 13. Epub ahead of print.

Ilami, G., Nespoulous, C., Huet, JC., Vartanian, N., Pernollet, JC., (1997) Characterization of BnD22, a drought-induced protein expressed in *Brassica napus* leaves. *Phytochemistry* 45, 1-8.

Itoh, R., Itoh, S., Sugawa, M., Oishi, O., Tabata, K., Okada, M., Nishimura, M., Yakushiji, E., (1982) Isolation of crystalline water-soluble chlorophyll proteins with different chlorophyll *a* and *b* contents from stems and leaves of *Lepidium virginicum. Plant Cell Physiol. 23, 557-560.*

Kamimura, Y., Mori, T., Yamasaki, T., Katoh, S., (1997) Isolation, properties and a possible function of a water-soluble chlorophyll *a*/*b*-protein from Brussels sprouts. *Plant Cell Physiol.* 38, 133-138.

Matile, P., Schellenberg, M., Vicentini, F., (1997) Localization of chlorophyllase in the chloroplast envelope. *Planta* 201, 96-99.

Murata, T., Ishikawa, C., (1981) Chemical, physicochemical and spectrophotometric properties of crystalline chlorophyll-protein complexes from *Lepidium virginicum* L. *Biochim. Biophys. Acta* 635, 341-347.

Murata, T., Murata, N., (1971) Water-soluble chlorophyll-proteins from *Brassica nigra* and *Lepidium virginicum. Carnegie Inst Wash Yearbook* 70, 504-507.

Murata, T., Toda, F., Uchino, K., Yakushiji, E., (1971) Water-soluble chlorophyll protein of *Brassica oleracea* var. *botrys* (cauliflower). *Biochim. Biophys. Acta* 245, 208-215.

Nishio, N., Satoh, H., (1997) A water-soluble chlorophyll protein in cauliflower may be identical to BnD22, a drought-induced, 22-kilodalton protein in rapeseed. *Plant Physiol.* 115, 841-846.

Reviron, MP., Vartanian, N., Sallantin, M., Huet, JC., Pernollet, JC., de Vienne, D., (1992) Characterization of a novel protein induced by progressive or rapid drought and salinity in *Brassica napus* leaves. *Plant Physiol.* 100, 1486-1493.

Sakuraba, Y., Schelbert, S., Park, SY., Han, SH., Lee, BD., Andrès, CB., Kessler, F., Hörtensteiner, S., Paek, NC., (2012) STAY-GREEN and chlorophyll catabolic enzymes interact at light-harvesting complex II for chlorophyll detoxification during leaf senescence in Arabidopsis. *Plant Cell* 24, 507-518.

Satoh, H., Nakayama, K., Okada, M., (1998) Molecular cloning and functional expression of a water-soluble chlorophyll protein, a putative carrier of chlorophyll molecules in cauliflower. *J. Biol. Chem.* 273, 30568-30575.

Satoh, H., Uchida, A., Nakayama, K., Okada, M., (2001) Water-soluble chlorophyll protein in Brassicaceae plants is a stress-induced chlorophyll-binding protein. *Plant Cell Physiol.* 42, 906-911.

Schmidt, K., Fufezan, C., Krieger-Liszkay, A., Satoh, H., Paulsen, H., (2003) Recombinant water-soluble chlorophyll protein from *Brassica oleracea* var. *Botrys* binds various chlorophyll derivatives. *Biochemistry* 42, 7427-7433.

Shinashi, K., Satoh, H., Uchida, A., Nakayama, K., Okada, M., Oonishi, I., (2000) Molecular characterization of a water-soluble chlorophyll protein from main veins of Japanese radish. *J. Plant Physiol.* 157, 255-262.

Takahashi, S., Ono, M., Uchida, A., Nakayama, K., Satoh, H., (2012a) Molecular cloning and functional expression of a water-soluble chlorophyll-binding protein from Japanese wild radish. *J. Plant Physiol.* In press.

Takahashi, S., Yanai, H., Nakamaru, Y., Uchida, A., Nakayama, K., Satoh, H., (2012b) Molecular cloning, characterization and analysis of the intracellular localization of a water-soluble chlorophyll-binding protein from Brussels sprouts (*Brassica oleracea* var. *gemmifera*). *Plant Cell Physiol.* 53, 879-891.

Takamiya, A., Yakushiji, E., (1976) Water-soluble chlorophyll protein. Tanpakushitsu Kakusan Koso Additional volume 386-396. [Article in Japanese]

Tanaka, R., Tanaka, A., (2007) Tetrapyrrole biosynthesis in higher plants. *Annu Rev. Plant Biol.* 58, 321-346.

Yakushiji, E., Uchino, K., Sugimura, Y., Shiratori, I., Takamiya, F., (1963) Isolation of water-soluble chlorophyll protein from the leaves of *Chenopodium Album. Biochim. Biophys. Acta* 75, 293-298.

Yamada, K., Hara-Nishimura, I., Nishimura, M. (2011) Unique defense strategy by the endoplasmic reticulum body in plants. *Plant Cell Physiol.* 52, 2039-2049.

In: Brassicaceae
Editor: Minglin Lang
ISBN: 978-1-62808-856-4

Chapter 6

THE PHYSIOLOGY, FUNCTIONAL GENOMICS, AND APPLIED ECOLOGY OF HEAVY METAL-TOLERANT BRASSICACEAE

Jillian E. Gall and Nishanta Rajakaruna*
College of the Atlantic, Bar Harbor, ME, US

ABSTRACT

Globally, \$25-50 billion is spent each year cleaning up sites contaminated with heavy metals. Because traditional cleanup methods such as incineration, chemical treatment, and excavation and removal are costly and can damage the environment, metal-hyperaccumulating plants (plants that accumulate >0.1% heavy metals in leaves or other tissues) may be a more cost-effective, less-intrusive option for remediating such sites. Members of the Brassicaceae comprise 25% of metal-hyperaccumulating species worldwide discovered to date and are potential candidates for phytoremediation technologies. Here we describe the diversity of metal-hyperaccumulating species in the Brassicaceae and discuss the physiological mechanisms of metal uptake and tolerance, the genetic basis for the metal tolerance mechanisms, ecological consequences of metal hyperaccumulation, and the role of the Brassicaceae species in remediating contaminated sites worldwide.

Keywords: Green technology, hyperaccumulation, metal tolerance, phytoremediation, ultramafic soils

INTRODUCTION

Heavy metals are highly reactive, toxic at low concentrations, and persist in the environment for years, posing severe risks to human and ecosystem health worldwide (Pilon-Smits, 2005; Neilson and Rajakaruna, 2012). Lead, for instance, may remain in the soil for

* Email: nrajakaruna@coa.edu.

150 to 5,000 years (Kumar *et al.*, 1995) and is known to cause cognitive dysfunction, neurobehavioral disorders, neurological damage, hypertension, and renal impairment in humans (Patrick, 2006). Although heavy metals are naturally present in the Earth's crust, human activities such as transportation, industrial manufacturing, commercial agriculture, mining, smelting, and military operations contribute largely to heavy metal pollution, releasing metals into the environment through waste disposal, runoff, and heavy metal-laden chemicals (Chaffai and Koyama, 2011; Pilon-Smits, 2005).

Vast areas of the world are contaminated with heavy metals (Ensley, 2000; Wuana and Okieimen, 2011). Traditional cleanup methods remove, incinerate or chemically treat contaminated soil, disrupting biotic communities and damaging the environment. These methods are also expensive, costing $25 to $50 billion worldwide annually (Neilson and Rajakaruna, 2011). The United States alone spends $6 to $8 billion each year cleaning up metal-contaminated sites (Tsao, 2003; Pilon-Smits, 2005), a steep investment that many developing nations cannot afford (Rajakaruna *et al.*, 2006).

Given the expense of conventional cleanup, there is much interest in seeking ecologically friendly, low-cost technologies to remove heavy metals from contaminated soils. One such alternative uses plants, a green technology known as phytoremediation (Krämer, 2005). Gaining popularity over the past few decades (Pilon-Smits and Freeman, 2006), phytoremediation utilizes metal hyperaccumulators, plants that can absorb, detoxify, and store high levels of heavy metals in their tissues. Hyperaccumulators take up high concentrations of heavy metals from the soil and translocate them into above-ground biomass at concentrations exceeding, in most cases, 0.1% of total dry leaf tissue mass (Baker *et al.*, 2000; Van der Ent *et al.*, 2012). The aboveground biomass can then be harvested and disposed of in a landfill or further processed for metal extraction (*i.e.* phytomining; Ghosh and Singh, 2005; Pilon-Smits, 2005).

Although some heavy metals, such as nickel (Ni), copper (Cu), iron (Fe), manganese (Mn), and zinc (Zn) regulate various biological processes in plants (Epstein and Bloom, 2004), when they occur in excess these metals may interact directly with biomolecules, disrupting critical biological processes (Kabata-Pendias, 2001; Chaffai and Koyama, 2011). Thus, most plants exclude metals at the roots by binding them to organic acids or ligands or storing them within vacuoles in the roots where they cannot interfere with important physiological processes (Hossain *et al.*, 2012).

Although metal-hyperaccumulators have the ability to detoxify and accumulate metals in their tissues, they do have limits to their extraordinary capacity to deal with metals, and the threshold for hyperaccumulation depends on the metal under consideration. Hyperaccumulators of cadmium (Cd), selenium (Se), and thallium (Tl) accumulate >100 $\mu g\ g^{-1}$ in their dry leaf tissue, cobalt (Co), chromium (Cr), and Cu accumulate >300 $\mu g\ g^{-1}$ in their dry leaf tissue, whereas lead (Pb), arsenic (As), antimony (At), and Ni accumulate >1,000 $\mu g\ g^{-1}$ in their dry leaf tissue. Hyperaccumulators of Zn accumulate >3,000 $\mu g\ g^{-1}$ while those of Mn accumulate >10,000 $\mu g\ g^{-1}$ in their dry leaf tissue (Reeves and Baker, 2000; Van der Ent *et al.* 2012). Although Se is a metalloid, we incorporate it into our discussion because it is a major environmental pollutant (Terry *et al.*, 2000). For recent reviews of metal-hyperaccumulation see Krämer (2010) and Van der ent *et al.* (2012).

Of the approximately 582 species of metal-hyperaccumulators from 50+ families of vascular plants worldwide, approximately 25% belong to the Brassicaceae, making it a model family for studying metal tolerance and hyperaccumulation (Rascio and Navari-Izzo, 2011;

Van der ent *et al.* 2012). In this chapter we introduce the metal-hyperaccumulating species in the Brassicaceae, outline the physiological mechanisms underlying their tolerance of heavy metals, and summarize the genetic basis for their remarkable physiology. We also discuss the ecological consequences of utilizing metal-hyperaccumulating species for phytoremediation, including the potential for metal transfer through trophic levels, the likelihood for invasiveness when employing non-native species, and the concerns of using genetically modified hyperaccumulators.

OVERVIEW OF METAL TOLERANCE IN THE BRASSICACEAE

The 93 documented species of metal-hyperaccumulating Brassicaceae (Table 1; Figure 1) provide substantial opportunity to study the physiological and genetic mechanisms behind metal tolerance and hyperaccumulation as well as the ecological implications of these mechanisms. Some of the most well-studied genera of hyperaccumulators in this family include *Arabidopsis*, *Brassica*, *Alyssum*, *Noccaea* (formerly *Thlaspi*), *Stanleya*, and *Streptanthus* (Bhargava *et al.*, 2012; Boyd *et al.*, 2009; Freeman *et al.*, 2010; Vamerali *et al.*, 2010; Verguggen *et al.*, 2009; Figure 1). Below we briefly introduce the most studied metal-hyperaccumulating Brassicaceae taxa.

Noccaea caerulescens (formerly *Thlaspi caerulescens*) is, perhaps, the most well-studied metal-hyperaccumulator (Milner and Kochian, 2008), accumulating up to 36,900 $\mu g\ g^{-1}$ Zn and 1800 $\mu g\ g^{-1}$ Cd without signs of toxicity (Bhargava *et al.*, 2012). Because *N. caerulescens*—like most other model taxa in the Brassicaceae—grows easily in the lab, it has been extensively studied, revealing several mechanisms for metal uptake, transport, and localization (Cosio *et al.*, 2004). However, its small biomass limits its potential as a candidate for phytoremediation (Bhargava *et al.*, 2012).

Arabidopsis thaliana, although not a naturally metal-accumulating species, is a popular model organism for plant-based research (Bevan and Walsh, 2005). *Arabidopsis thaliana*'s genome is mapped (Weigel and Mott, 2009) and its sequence is very similar to its metal-accumulating congener *A. halleri* (Becher *et al.*, 2004; Meyer and Verbruggen, 2012; Weber *et al.*, 2004). For this reason, both *A. thaliana* and *A. halleri* are commonly used to study the genetic basis for metal tolerance and hyperaccumulation (Bevan and Walsh, 2005; Cho *et al.*, 2003; Chaffai and Koyama, 2011; Courbot *et al.*, 2007; Hanikenne *et al.*, 2008).

A common condiment crop in North America and Europe, *Brassica juncea* (Indian mustard) is a popular choice for phytoremediation (Lim *et al.*, 2004; Neilson and Rajakaruna, 2012). Although not a hyperaccumulator, with the ability to accumulate Cd, Zn, Se, and Pb and a biomass at least 10-fold greater than that of *N. caerulescens*, *B. juncea* has been used with success in several phytoremediation studies and trials (Bhargava *et al.*, 2012; Szczygłowska *et al.*, 2011; Warwick, 2011).

The molecular mechanisms responsible for selenium (Se) tolerance and hyperaccumulation have been investigated in the Se hyperaccumulator *Stanleya pinnata* by comparing it with its Se-tolerant congener, *S. albescens*, using a combination of physiological, structural, genomic, and biochemical approaches (Freeman *et al.*, 2010). Additionally, the ecological functions and implications of Se hyperaccumulation in *Stanleya* and other plants (El Mehdawi and Pilon-Smits, 2012) and the potential for Se phytoremediation have also been investigated (Bañuelos, 2001).

Table 1. Brassicaceae species known to hyperaccumulate heavy metals based on an extensive review of the literature and personal communication with Dr. Roger D. Reeves. Nomenclature follows International Plant Names Index [website (http://www.ipni.org/index.html); accessed Oct 2012]. The two taxa with * have questionable nomenclature in light of recent taxonomic revisions in the group (personal communication Ihsan A. Al-Shehbaz, Senior Curator, Missouri Botanical Garden, USA)

Species	Metal Hyperaccumulated	Reference
Aethionema spicatum Post	Ni	Reeves *et al*., 2001
Alyssum akamasicum Burtt	Ni	Brooks *et al*., 1979
A. alpestre L.	Ni	Brooks and Radford, 1978
A. anatolicum Hausskn. ex Nyár.	Ni	Brooks *et al*., 1979
A. argenteum All.	Ni	Brooks and Radford, 1978
A. baldaccii Vierh. ex Nyár.	Ni	Brooks and Radford, 1978
A. bertolonii Desv. subsp. *scutarinum* Nyár.	Ni	Minguzzi and Vergnano, 1948; Brooks *et al*., 1979; Reeves *et al*., 1983
A. bracteatum Boiss. & Buhse	Ni	Ghaderian *et al*., 2007a
A. callichroum Boiss. & Bal.	Ni	Brooks *et al*., 1979
A. caricum T.R. Dudley & Hub.-Mor.	Ni	Brooks *et al*., 1979
A. cassium Boiss.	Ni	Brooks *et al*., 1979
A. chalcidicum Janka	Ni	Brooks and Radford, 1978
A. cholorocarpum Hausskn	Ni	Brooks and Radford, 1978
A. chondrogynum Burtt	Ni	Brooks *et al*., 1979
A. cilicicum Boiss. & Bal.	Ni	Brooks *et al*., 1979
A. condensatum Boiss.	Ni	Brooks *et al*., 1979
A. constellatum Boiss.	Ni	Brooks *et al*., 1979
A. corsicum Duby	Ni	Brooks *et al*., 1979
A. crenulatum Boiss.	Ni	Brooks *et al*., 1979
A. cypricum Nyár.	Ni	Brooks *et al*., 1979
A. davisianum T.R. Dudley	Ni	Brooks *et al*., 1979
A. discolor T.R.Dudley & Hub-Mor.	Ni	Brooks *et al*., 1979
A. dubertretii Gomb.	Ni	Brooks *et al*., 1979
A. dudleyi N. Adigüzel & R.D. Reeves	Ni	Adigüzel and Reeves, 2002
A. eriophyllum Boiss. & Hausskn.	Ni	Brooks *et al*., 1979
A. euboeum Halácsy	Ni	Brooks and Radford, 1978
A. floribundum Boiss. & Bal.	Ni	Brooks *et al*., 1979
A. giosnanum Nyár.	Ni	Brooks *et al*., 1979
A. heldreichii Hausskn	Ni	Brooks and Radford, 1978
A. huber-morathii T.R.Dudley	Ni	Brooks *et al*., 1979
A. inflatum Nyár.	Ni	Ghaderian *et al*., 2007b
A. lesbiacum Candargy (Rech. f.)	Ni	Brooks *et al*., 1979
A. longistylum Grossh.	Ni	Ghaderian *et al*., 2007b
A. markgrafi O.E. Schulz	Ni	Brooks and Radford, 1978
A. masmenaeum Boiss.	Ni	Brooks *et al*., 1979
A. murale Waldst. & Kit. subsp. *haradjianii* (Rech.) T.R.Dudley subsp. *pichleri* (Velen.) Stoj. & Stef	Ni	Doksopulo, 1961; Reeves *et al*., 2001; Reeves and Adıgüzel, 2008; Reeves *et al*., 1983

Species	Metal Hyperaccumulated	Reference
A. obovatum (C.A. Meyer) Turcz.	Ni	Brooks *et al.*, 1979
A. oxycarpum Boiss. & Bal.	Ni	Brooks *et al.*, 1979
A. pateri Nyár.	Ni	Reeves and Adigüzel, 2008
A. peltarioides Boiss.	Ni	Reeves and Adigüzel, 2008
subsp. *virgatiforme* (Nyár.) T.R.Dudley		Reeves *et al.*, 1983; Reeves and Adıgüzel, 2008
A. penjwinensis T.R. Dudley	Ni	Brooks *et al.*, 1979
A. pinifolium (Nyár.) T.R.Dudley	Ni	Brooks *et al.*, 1979
A. pintodasilvae T.R.Dudley	Ni	Gonçalves *et al.*, 2007
A. pterocarpum T.R.Dudley	Ni	Brooks *et al.*, 1979
A. robertianum Bernard ex Gren. & Godr.	Ni	Brooks and Radford, 1978
A. samariferum Boiss. & Hausskn.	Ni	Brooks *et al.*, 1979
A. serpyllifolium Desf.	Ni	Brooks and Radford, 1978
subsp. *lusitanicum* T.R. Dudley & Pinto da Silva		Brooks *et al.*, 1981
subsp. *malacitanum* Rivas Goday		
A. sibiricum Willd.	Ni	Brooks *et al.*, 1979
A. singarense Boiss. & Hausskn.	Ni	Brooks *et al.*, 1979
A. smolikanum Nyár.	Ni	Brooks and Radford, 1978
A. syriacum Nyár.	Ni	Brooks *et al.*, 1979
A. tenium Halácsy	Ni	Brooks and Radford, 1978
A. trapeziforme Bornm. Ex Nyár.	Ni	Brooks *et al.*, 1979
A. troodii Boiss.	Ni	Brooks *et al.*, 1979
A. virgatum Nyár.	Ni	Brooks *et al.*, 1979
Arabidopsis halleri (L.) O'Kane & Al-Shehbaz	Zn Cd	Ernst, 1968 Zhao *et al.*, 2000; Bert *et al.*, 2002
Arabis gemmifera (Matsum.) Makino	Zn	Kubota and Takenaka, 2003
A. paniculata Franch.	Zn Cd Pb	Tang *et al.*, 2009
Bornmuellera baldacci (Degen Heywood)	Ni	Reeves *et al.*, 1983
subsp. *baldacci*		Reeves *et al.*, 1983
subsp. *markgrafi* (Schulz ex Markgraf) Dudley		Reeves *et al.*, 1983
subsp. *rechingeri* Greuter		
B. glabrescens (Boiss. & Bal.) Cullen & T.R. Dudley	Ni	Reeves *et al.*, 1983
B. kiyakii Aytaç & A.Aksoy	Ni	Reeves & Adıgüzel 2009
B. tymphaea (Hausskn.) Hausskn.	Ni	Reeves *et al.*, 1983
B. x petri Greuter, Charpin & Dittrich	Ni	Reeves *et al.*, 1983
Brassica oleracea L.	Ti	Al-Najer *et al.*, 2005
Cardamine resedifolia L.	Ni	Vergnano Gambi and Gabbrielli, 1979
Iberis intermedia Guers.	Tl	Leblanc *et al.*, 1999
Leptoplax emarginata (Boiss.) O.E. Schulz	Ni	Reeves *et al.*, 1980
Masmenia rosularis (Boiss. & Bal.) F.K. Meyer	Ni	Reeves, 1988
Microthlaspi perfoliatum (L.) F.K.Mey. (as *T. perfoliatum* L.)	Ni	Reeves *et al.*, 2001
Noccaea caerulescens (J.Presl & C.Presl) F.K.Mey. (as *T. caerulescens* J.Presl & C.Presl)	Zn Ni Cd	Sachs, 1865; Ernst, 1966, 1968, 1974; Reeves and Brooks, 1983; Lombi *et al.*, 2000; Escarré *et al.*, 2000; Reeves *et al.*, 2001

Table 1. Continued

Species	Metal Hyperaccumulated	Reference
N. cariensis (Carlström) Parolly, Nordt & Aytaç (as *T. cariense* A. Carlström)	Ni	Reeves *et al.*, 2001
N. cepaefolia (Wulfen) Rchb. (as *T. rotundifolium* (L.) Gaudin subsp. *cepaeifolium* (Wulfen) Rouy & Fouc.)	Zn	Rascio, 1977; Reeves and Brooks, 1983
N. cochleariforme (as *T. japonicum* H.Boissieu)	Ni	Reeves 1988; Mizuno *et al.*, 2003, 2005
N. epirota (Halácsy) F.K.Mey. (as *T. epirotum* Halácsy)	Ni	Reeves and Brooks, 1983
N. fendleri subsp. *californica* (S. Watson) Al-Shehbaz & M. Koch (*T. montanum* var. *californicum* (Watson) P.K.Holmgren) *N. fendleri* subsp. *fendleri* (A. Gray) Holub (as *Thlaspi montanum* var. *fendleri* (A.Gray) P.K.Holmgren) *N. fendleri* subsp. *siskiyouensis* (P.K. Holmgren) Al-Shehbaz & M. Koch (as *T. montanum* var. *siskiyouense* P.K.Holmgren)	Ni	Reeves *et al.*, 1983
N. goesingense (Halácsy) F.K.Mey. (as *T. goesingense* Halácsy)	Ni	Reeves and Baker, 1984
N. graeca (Jord.) F.K.Mey. (as *T. graecum* Jord.)	Ni	Reeves and Brooks, 1983
N. kovatsii (Heuff.) F.K.Mey. (as *T. kovatsii* Heuffel)	Ni	Bani *et al.*, 2010
N. ochroleuca (Boiss. & Heldr.) F.K.Mey. (as *T. ochroleucum* Boiss. & Heldr.)	Ni	Reeves and Brooks, 1983
N. praecox (Wuljen) F.K.Mey. (as *T. praecox* Wulfen in Jacq.)	Ni	Bani *et al.*, 2010
N. pindica (Hausskn.) Holub (as *T. pindicum Hausskn.*)	Ni	Taylor and Macnair, 2006
N. tymphaea (Hausskn.) F.K.Mey. (as *T. tymphaeum* Hausskn. and *T. goesingense* Halácsy)	Ni	Reeves and Brooks, 1983
Pseudosempervivum aucheri (Boiss.) Pobed. (as *Cochlearia aucheri* Boiss.)	Ni	Reeves, 1988
P. semprevivum Boiss. & Bal.) Pobed. (as *Cochlearia sempervivum* Boiss. & Balansa)	Ni	Reeves, 1988
Raparia bulbosa (Spruner) F.K.Mey. (as *Thlaspi bulbosum* Spruner)	Ni	Reeves and Brooks, 1983
Stanleya bipinnata Greene		Moxon *et al.*, 1950; Rosenfeld and Beath 1964
S. pinnata (Pursh) Britton	Se	Rosenfeld and Beath, 1964
Streptanthus polygaloides A. Gray	Ni	Reeves *et al.*, 1981
**Thlaspi jaubertii* Hedge	Ni	Reeves and Brooks, 1983
**T. rosulare* Boiss. & Bal.	Ni	Reeves and Adigüzel, 2008
Thlaspiceras eigii (Zohary) F.K.Mey. subsp. *samuelssonii* F.K.Mey. (as *Thlaspi eigii* (Zohary) Greuter & Burdet subsp. *samuelssonii* (F.K.Mey.) Greuter & Burdet)	Ni	Reeves and Adigüzel, 2008
T. elegans (Boiss.) F.K.Mey. (as *T. elegans* Boiss.)	Ni	Reeves and Brooks, 1983
T. oxyceras (Boiss.) F.K.Mey. (as *T. oxyceras* (Boiss.) Hedge)	Ni	Reeves and Adigüzel, 2008

Figure 1. Metal-hyperaccumulating species. A, *Alyssum bertolonii* (Brassicaceae) was first reported by Caesalpino (1583) as confined to Ni-rich serpentine outcrops near Florence Italy. Minguzzi and Vergnano (1948) discovered that this plant had an extraordinarily high Ni content of about 10,000 ppm [1%] in dried matter which translated to well over 10% of Ni in the ash. Photo Credit: Dr. A. J. M. Baker; B, *Noccaea caerulescens* (Brassicaceae) hyperaccumulates Zn, Cd, and Ni and is the 'lab rat' for metal-hyperaccumulating research. Photo Credit: Dr. A. J. M. Baker; C, *Streptanthus polygaloides* (Brassicaceae), a Californian serpentine endemic, is one of only two Ni-hyperaccumulating species found in North America. Photo Credit: Dr. Robert S. Boyd; D, *Stanleya pinnata* (Brassicaceae) is a Se-hyperaccumulating perennial species native to Southwestern United States. Photo Credit: Malia Volke; E, *Bornmuellera tymphaea* (Brassicaceae) is a Ni-hyperaccumulating species native to serpentine soils in Greece. Photo Credit: Dr. Roger D. Reeves; F, *Alyssum murale* (Brassicaceae) is a widespread and polymorphic Ni-hyperaccumulating species native to eastern Mediterranean Europe, Turkey and adjacent parts of SW Asia. Photo Credit: Dr. Roger D. Reeves.

Endemic to ultramafic (serpentine) soils along the western side of the Sierra Nevada in California, *Streptanthus polygaloides* is a small annual that hyperaccumulates Ni (Pope *et al.*, 2013) in concentrations ranging from 1100 to 16,400 $\mu g\ g^{-1}$ dry mass in its leaves, stems, roots, flowers, and fruits (Reeves *et al.*, 1981). *Streptanthus polygaloides* is one of two native Ni hyperaccumulators confirmed from continental North America (Reeves *et al.*, 1981; Boyd *et al.*, 2009); the other, *Noccaea fendleri*, is also from the Brassicaceae (O'Dell and Rajakaruna, 2011). The relatively small aboveground biomass of these species makes *S. polygaloides* and *N. fendleri* poor candidates for phytoremediation. However, *S. polygaloides* has been investigated for its potential for phytoremediation and phytomining (Anderson *et al.*, 1999). Additionally, its ecology (Jhee *et al.*, 2005; Boyd *et al.*, 2009; Pope *et al.*, 2013), including the role it may play in the transfer of metals through the food web (Wall and Boyd, 2002), has also been studied. Wall and Boyd (2002) discovered an insect, *Melanotrichus boydi* (Hemiptera: Miridae), which is monophagous on *S. polygaloides*. They reported that *M. boydi* accumulates Ni up to nearly 800 $\mu g\ g^{-1}$ of dry tissue, raising concerns about the potential for metals to transfer from hyperaccumulating plants to the insects that feed on them (Peralta-Videa *et al.*, 2009).

PHYSIOLOGICAL MECHANISMS OF METAL TOLERANCE

In the soil, metal cations are bound to negatively charged particles such as clay and organic matter. Hyperaccumulators take up metals (Figure 2a) only after the cations have detached from these soil particles due to mass ion effect and become bioavailable in the soil solution (Neilson and Rajakaruna, 2012). The bioavailability of a metal depends on the interaction of various physical, chemical, and biological processes within the soil (Maestri *et al.*, 2010; Jabeen *et al.*, 2009). At low pH, certain metals such as Cd, Cu, Hg, and Pb become more available for plant uptake (Blaylock and Huang, 2000). Some hyperaccumulators release protons or metal-chelators such as mugenic and aveic acids (Jabeen *et al.*, 2009) which acidify the rhizosphere, freeing metals into the soil solution (Salt *et al.*, 1994). Soil bacteria may also release a number of compounds into the soil such as antibiotics, antifungals, organic acids, hormones, and metal chelators, which may increase the bioavailability of metals (Xiong *et al.*, 2008). Puschenreiter *et al.* (2005) found that metals were more bioavailable in the rhizosphere of hyperaccumulators than in non-hyperaccumulators, suggesting that hyperaccumulators may actively alter their rhizospheric environment to increase the availability of metals (Neilson and Rajakaruna, 2012). The root structure of hyperaccumulators also appears to differ from that of non-hyperaccumulators. Mench *et al.* (2009) identified a zone in the roots of *N. caerulescens,* external to the endodermis near the root tip, with thickened inner tangential cell walls which may assist metal uptake and transport.

When bioavailable metals come into contact with roots (either through diffusion or bulk flow), they enter the root apoplast. Metal ions may remain in the apoplast, traveling passively from cell wall to cell wall, or they may cross the plasma membrane into the symplast through a number of embedded ion transport proteins (Figure 2b) including pumps, channels, or carriers (Jabeen *et al.*, 2009; Salt *et al.*, 1995). In addition to regulating metal ions within the cell, ion transporter proteins are critical for metal uptake. Plants have several classes of metal-transporters: the heavy metal (or CPX-type) ATPases, the natural resistance-associated macrophage-proteins (Nramp), cation-diffusion facilitator (CDF) proteins, zinc-iron permeases (ZIP), cation exchangers (CAXs), and copper transporters (COPTs) (Chaffai and Koyama, 2011; Jabeen *et al.*, 2009). ZIP family proteins have been shown to be particularly important for metal uptake in *N. caerulescens* (Chaffai and Koyama, 2011; Maestri *et al.*, 2010).

Once a metal ion has entered the symplast of a root cell via a metal transporter, it can either be sequestered into the root vacuoles or transported to the leaves via the xylem. In either situation, chelators must bind to the metals to protect the internal environment of the cell from metal-induced damage (Memon and Schroder, 2008). Chelators include organic acids such as citrate, malate, and malonate, and proteins such as histidine, metallothioneins, and phytochelatins (Maestri *et al.*, 2010; Pilon-Smits and Pilon, 2002). Metallothioneins and phytochelatins are thiol-rich ligands which donate electrons more readily than oxygen, thus forming stable complexes with first-row transition metals (Baker *et al.*, 2000). Organic acids and ligands bind metals differentially throughout the plant. Verbruggen *et al.* (2009) found that histidine accompanied Zn in the roots of *N. caerulescens*, whereas organic acids accompanied Zn in the shoots. They also found Cd bound to sulfur ligands in the leaves of *N. caerulescens*. Organic acids are also prevalent both in the cytosol and in the acidic vacuoles of root and leaf cells. Verbruggen *et al.* (2009) observed Zn-malate complexes in epidermal cell vacuoles of *N. caerulescens* and in mesophyll vacuoles of *A. halleri*. Malate also transports Zn from the cytosol to the vacuole, where it transfers Zn to sulfur-containing mustard oils before being transported back to the cytosol to retrieve more Zn (Baker *et al.*, 2000).

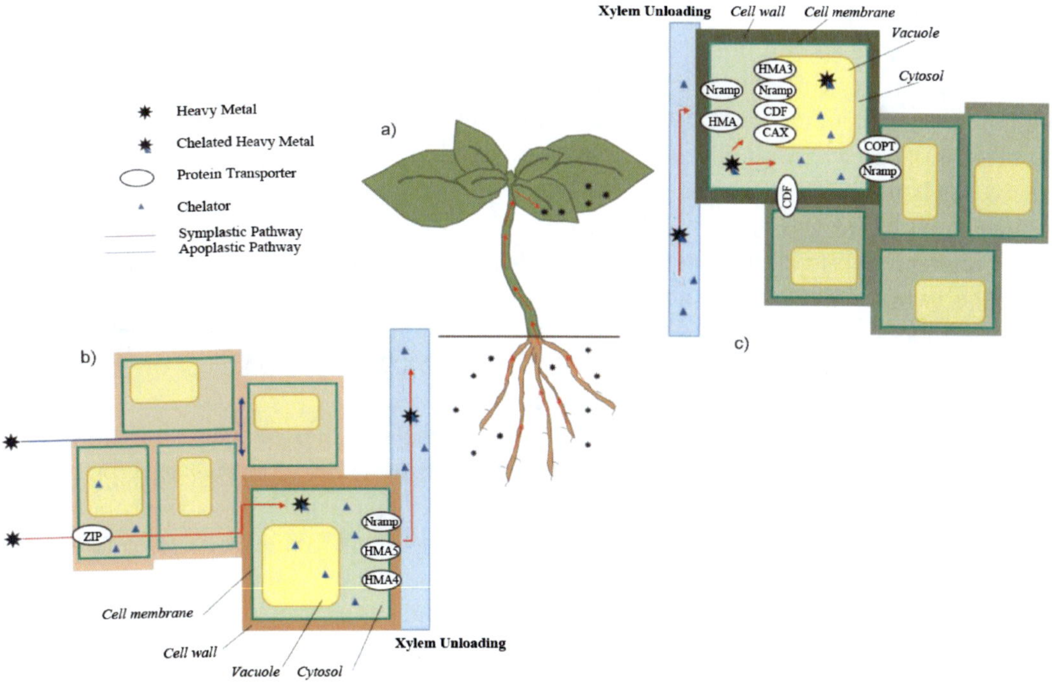

Figure 2. Mechanism of metal hyperaccumulation and tolerance. a) Translocation of heavy metals from roots to leaves in hyperaccumulator plants b) Heavy metals enter the roots and travel cell to cell through an apoplastic or symplastic pathway. In the symplastic pathway, heavy metals pass through specialized transporter proteins and are chelated, eventually loading into the xylem. c) Heavy metals in the xylem unload into leaf cells and either enter the vacuole or move to neighboring cells. Adapted from Maestri *et al.*, 2010.

For metals to end up in the leaves of a hyperaccumulator, they must be loaded into the xylem, moved up the shoot, and deposited in the vacuoles of the leaf cells (Figure 2). Such movement requires passing through at least three plasma membranes: the plasma membrane of the root cell, the plasma membrane of the leaf cell, and the tonoplast of the leaf cell vacuole. The rate at which metals move through the xylem depends on the metal concentration in the root, with higher concentrations of metals in the roots resulting in faster loading into the xylem (Jabeen *et al.*, 2009). ATPase and Nramp class transporters are particularly important for xylem loading; many members of ATPase have been identified in both *A. halleri* and *A. thaliana* (Maestri *et al.*, 2010; Jabeen *et al.*, 2009).

Once in the xylem, low molecular weight chelators (such as malate, citrate, phytochelatins, and free histidine) bind to metal ions and transpiration pulls the complexed metals up the xlyem to the leaves (Jabeen *et al.*, 2009). High levels of histidine in roots of hyperaccumulators may also increase translocation of metals from roots to leaves. Richau *et al.* (2009) found that as concentrations of histidine increased in roots, Ni-histidine complexes decreased in root vacuoles. They also found that the concentration of histidine in roots of *N. caerulescens* (reported as *T. caerulescens*) was 10-fold higher than in the non-metal-hyperaccumulating congener *N. arvense* (reported as *T. arvense*), although the amount of histidine found in the leaves was only slightly greater in *N. caerulescens*. Richau *et al.* (2009) also exposed three hyperaccumulating *Alyssum* species to Ni and reported an increase of histidine in the xylem sap of all three species, a phenomenon not observed in the non-hyperaccumulating species of the Brassicaceae.

To exit the xylem, metals must pass through the leaf cell wall and cell membrane (Figure 2c), a process regulated by ATPase and Nramp proteins (Jabeen *et al.*, 2009). Once in the cytosol of the leaf, proteins such as ATPases and phytochelatins transport metals to the vacuoles where they are bound to organic acids or anthocyanins and are stored until senescence (Chaffai and Koyama, 2011; Pilon-Smits and Pilon, 2002). Hyperaccumulators often preferentially store more metals in shoot vacuoles than root vacuoles, with the opposite being the case for non-hyperaccumulators. In the hyperaccumulator *N. caerulescens*, Ni is higher in shoot tonoplast vesicles than in root tonoplast vesicles with the opposite pattern in the non-hyperaccumulator, *N. arvense* (Richau *et al.*, 2009). These findings suggest that the mechanisms of metal tolerance are species-specific.

Although hyperaccumulators sequester metals in the vacuoles of their leaves, the exact location of this sequestration within leaves varies by species. Broadhurst *et al.* (2004) found that in five *Alyssum* species, Ni was stored in either leaf epidermal cell vacuoles or in basal portions of stellate trichomes, reporting that the Ni in trichomes comprised a remarkable 15% to 20% of the plant's dry weight. However, Ghasemi *et al.* (2009) did not find any more Ni in trichomes than in shoots of *A. inflatum*, a Ni-hyperaccumulating species of *Alyssum* native to Iran. However, after immersing whole *A. inflatum* leaves in Ni-indicating dimethylglyoxime (DMG), staining of trichomes increased with Ni exposure, showing that trichomes of this species are capable of accumulating high levels of Ni.

Seasonality has also been shown to affect where hyperaccumulators sequester metals (Bidar *et al.*, 2008). Galeas *et al.* (2006) observed that Se was transported from roots to young leaves of *Stanleya pinnata* in the spring, from old leaves to flowers in the summer, and back to the roots in the fall. Such seasonal patterns may complicate phytoremediation efforts, especially if a phytoremediating plant is harvested during a season when metals are not present in above-ground biomass.

GENETICS OF METAL TOLERANCE

Hyperaccumulation often results from the overexpression of genes which code for specialized protein transporters and chelators (Chaffai and Koyama, 2011; Rascio and Navari-Izzo, 2011; Maestri *et al.*, 2010; Verbruggen *et al.*, 2009; Jabeen *et al.*, 2009; Cobbett and Goldsbrough, 2002; Figure 2b, c). Below we describe what is known about the genetic basis of metal transporters and chelators from studies conducted on species of the Brassicaceae. Genes currently known to code for metal transporters and chelators in the Brassicaceae are listed in Table 2.

Table 2. The metal transporter genes characterized from Brassicaceae species

Species	Gene	Metal Transported	Reference
Arabidopsis halleri	hma4	Cd	Courbot *et al.* 2007
	nas2, nas3	Zn	Talke *et al.* 2006
A. thaliana	zip1-12	Zn	Weber *et al.*, 2004; Roosens *et al.*, (2008a,b)
	irt1	Fe	Kerkeb *et al.*, 2008
	mtp1	Zn	Kawachi *et al.*, 2008
	hma3	Co, Zn, Cd, Pb	Morel *et al.*, 2008
	copt1	Cu	Sancenon *et al.*, 2004; Andres-Colas *et al.*, 2010
	yls2	Fe, Cu	DiDonato *et al.*, 2004
Noccaea caerulescens (as *Thlaspi caerulescens*)	znt1-2	Zn	van de Mortel *et al.*, 2006
	irt1-2	Fe	Schikora *et al.*, 2006; Plaza *et al.*, 2007
	ysl3	Fe, Ni	Gendre *et al.* 2006
N. cochleariforme (as *Thlaspi japonicum*)	nramp4	Fe	Mizuno *et al.*, 2005
N. goesingense (as *Thlaspi goesingense*)	mtp1	Zn, Ni	Kim *et al.*, 2004

Transporters

The ZIP protein transporter family was one of the first metal transporter groups identified in plants. ZIP transporters take up cations, particularly Zn and Fe, in different plant species including *Arabidopsis* (Rascio and Navari-Izzo, 2011). In *A. thaliana*, 15 genes have been documented to code for transporters of various metals (Chaffai and Koyama, 2011). Interestingly, non-hyperaccumulating *Arabidopsis* only expressed ZIP transporters when deficient in Zn, whereas hyperaccumulating *Arabidopsis* expressed ZIP transporters independent of Zn levels (Verbruggen *et al.*, 2009). This suggests that ZIP transporters are constitutively expressed in the hyperaccumulator and not in the non-hyperaccumulator. Similarly, both hyperaccumulating *Noccaea caerulescens* and metal-excluding *Noccaea arvense* (as *Thlaspi caerulescens* and *T. arvense*) have a ZIP protein with similar affinities for Zn. Because the affinities for Zn do not differ, this protein is likely expressed at higher rates in hyperaccumulating *N. caerulescens* than in *N. arvense*, resulting in a greater number of membrane proteins that transport Zn (Lasat and Kochian, 2000).

ATPase protein transporters use ATP to transport cations within and between cells, especially between root and shoot cells. In hyperaccumulating *A. halleri*, HMA4 and HMA5 transporters move metal ions from root to xylem cells, increasing tolerance of metals in the roots (Chaffai and Koyama, 2011). When AtHMA4 in *A. thaliana* was artificially overexpressed, higher levels of Zn and Cd were translocated from the root to the shoot (Verret *et al.*, 2004). In *A. halleri*, the HMA4 gene is consistently overexpressed, accounting for the greater Cd-tolerance of *A. halleri* (Hanikenne *et al.* 2008). In contrast with HMA4 and

HMA5, which are predominantly expressed in the roots, HMA3 transporters reside in the vacuolar tonoplast and AtHMA3 regulates Zn levels in the vacuole in shoots of *A. thaliana* (Gravot *et al.*, 2004).

Similar to ATPase proteins, COPT proteins transport Cu^{2+} within and between cells. Because COPT proteins deal exclusively with Cu^{2+}, they have a higher specificity than some ZIP family transporters. Non-hyperaccumulating *Arabidopsis thaliana* has five genes that encode COPT proteins. Over-expression of COPT1 proteins in *Arabidopsis* appears to increase Cu tolerance in the roots (Kobayashi *et al.*, 2008; Puig *et al.*, 2007).

The Nramp family of transporter proteins contains at least seven members, five of which have been characterized. *Arabidopsis thaliana* encodes six Nramp-like proteins located in different parts of the cell which regulate metal homeostasis. AtNRAMP3 is located in the tonoplast and is responsible for transporting Fe, Cd, Mn, and Zn between the vacuole and the cytosol (Chaffai and Koyama, 2011; Thomine *et al.*, 2003). AtNRAMP3 also appears to be controlled by the presence of Fe. In Fe-sufficient conditions, over-expressing or disrupting AtNRAMP3 does not change the metal content of the cell. However, in Fe-starved conditions, overexpressing AtNRAMP3 decreases overall plant Zn and Mn concentrations, while disrupting AtNRAMP3 invokes the opposite response (Thomine *et al.*, 2003). This suggests that AtNRAMP3's function is tightly linked with Fe and together they regulate concentrations of Mn and Zn within the cell. AtNRAMP4 and AtNRAMP6 proteins transport ions within the cell, with AtNRAMP4 responsible for transporting Fe, Mn, Cd, and Zn and AtNRAMP6 responsible for transporting Cd and regulating the distribution and availability of Fe and Mn.

Embedded primarily in tonoplast and cell plasma membranes, CDF and CAX transporter proteins regulate the concentration of metal ions in the cytoplasm of hyperaccumulators. Divided into Mn^{2+}, Zn^{2+}, and Fe^{2+}/Zn^{2+} groups, CDF transporters pump H^{+} or K^{+} either outside of the cell or into the vacuole, regulating the concentration of heavy metals in the cytoplasm (Chao and Fu, 2004). Similarly, CAX transporters pump H^{+} or Na^{+} outside of the cell or into the vacuole, regulating the concentration of heavy metals in the cytoplasm (Hall and Williams, 2003).

Chelators

Cytosine-rich and with a low molecular weight, metallothionein chelators form mercaptide bonds with a range of metals (Maestri *et al.*, 2010). Metallothioneins comprise four subfamilies: MT1, MT2, MT3, and MT4. The expression of each group varies between hyperaccumulators and non-hyperaccumulaters. When non-hyperaccumulators such as *A. thaliana* are exposed to Cd, Cu, or Zn, they express MT1a and MT1b at high levels in the roots, whereas in hyperaccumulators these genes are expressed at higher levels in the leaves (Maestri *et al.*, 2010). Overexpression of MT2 is associated with Cu-tolerance in *A. halleri* and *N. caerulescens*. In *N. caerulescens* metallothionein MT3 is also associated with Cu-tolerance; however, this transporter is expressed only when Cd is present. MT4 chelators have been shown to maintain Cu homeostasis in the seeds of *A. thaliana* (Maestri *et al.*, 2010).

Like metallothioneins, phytochelatins (PCs) detoxify heavy metals in a number of species, binding with metals in the cytosol of the roots and leaves to form stable heavy metal complexes that are deposited in the vacuoles (Cobbett, 2000). These cytesine-rich phytochelatins are synthesized by PC synthase, an enzyme that binds to protein substrate glutathione (GSH). Several studies confirm that *Arabidopsis* mutants deficient in GSH are also deficient in PC synthase (Lee *et al.*, 2003). In *A. thaliana*, the AtPCS1 and CAD1 genes encode PC synthase, and CAD2 encodes GSH (Cobbett and Goldsbrough, 2002). PCs have been shown to play a major role in Cd detoxification, as mutant lines of *A. thaliana* with defective PC synthase are intolerant of Cd. PC production can be induced in plants and cultured cells exposed to metal ions, particularly Cd. To increase the level of these metal-binding peptides and enhance heavy metal tolerance, PC synthase genes from *A. thaliana* (Ha *et al.*, 1999; Vatamaniuk *et al.*, 1999) and *Brassica juncea* (Heiss *et al.*, 2003) have been inserted and overexpressed in a number of plant species (Lee *et al.*, 2011).

Ecological Implications

Hypotheses for Metal Hyperaccumulation

Many authors have hypothesized about the reasons for metal hyperaccumulation in plants (Boyd, 2010). However, only a few of these hypotheses have been tested experimentally (Boyd and Martens, 1998; Boyd, 2007). Some of the reasons hypothesized for metal-hyperaccumulation include the ability to tolerate the metal and dispose it from the plant body (Baker, 1981), drought resistance (Baker and Walker, 1989), elemental allelopathy (Boyd and Jaffré, 2001), inadvertent uptake (Cole, 1973), and the most widely tested, pathogen/herbivore defense (Reeves *et al.*, 1981). Species of Brassicaceae have often been utilized to shed light on these various hypotheses.

In the *metal tolerance hypothesis*, it is believed that plants sequester metals in their cell walls and vacuoles to avoid toxicity, keeping metals away from metabolically active sites in the cell (Kruckeberg and Reeves, 1995). The *disposal hypothesis* suggests that plants either store metals in tissues which are about to be shed by the plant or in the epidermal cells of the leaf where the metals may be washed out by rainfall (Farago and Cole, 1988). Similarly, the *elemental allelopathy hypothesis* suggests that hyperaccumulators shed metal-laden leaves to increase the metal concentration of the soil (Boyd and Jaffré, 2001), thus keeping metal-intolerant competitors at bay. The *drought resistance hypothesis* (Baker and Walker, 1989) suggests that plants may use metals to prevent drought by increasing the concentration of ions within the roots, thus creating negative water potential which draws water into the plant. In the *Mimulus guttatus* complex (Phrymaceae), drought tolerance also appears to provide tolerance to metal-enriched serpentine soils (Hughes *et al.*, 2001). The *inadvertent uptake hypothesis* assumes that hyperaccumulation is a by-product of another adaptive function (Boyd and Martens, 1992). The most experimentally tested and commonly accepted hypothesis for metal hyperaccumulation is the *elemental defense hypothesis*, suggesting that metal sequestration in the leaf tissue defends plants against insect herbivory and infection by pathogens (Strauss and Boyd, 2011; Boyd and Martens, 1998).

Martens and Boyd (2002) tested the effects of herbivory on the Ni-hyperaccumulating *S. polygaloides* and found that elevated levels of Ni did not always prevent herbivore attack. They suggested, instead, that Ni-hyperaccumulation may only defend plants against some herbivores and that herbivores can overcome these plant defenses through the evolution of metal-tolerance or detoxification mechanisms. These findings are similar to those of Wall and

Boyd (2002) who examined the arthropods associated with the flora of an ultramafic site in the Red Hills of California. They found elevated levels of Ni in arthropods of serpentine sites relative to levels of Ni in arthropods of non-serpentine sites and discovered the insect *M. boydi* to be monophagous on *S. polygaloides.* Boyd defined this insect as a high-Ni insect (Boyd, 2009), accumulating Ni at levels up to 777 μg g^{-1} (the minimum requirement for a high-Ni insect is 500 μg Ni g^{-1} dry tissue). Whereas the *elemental defense hypothesis* suggests that *S. polygaloides* may hyperaccumulate Ni to deter herbivores, *M. boydi* may have specialized on *S. polygaloides* to potentially deter predators.

To date, Ni is the most explored element in terms of the *elemental defense hypothesis*, with most studies focusing on Ni-hyperaccumulators (Strauss and Boyd, 2011). Boyd (2007), in a recent review of the elemental defense hypothesis, calls for further experiments with a wider array of metals (see Barillas *et al.*, 2011 for studies on Se). Boyd (2004) noted that elemental defense may also occur in plants which accumulate metals below the threshold for hyperaccumulation and that studies should be directed at plant-biota interactions even in plants accumulating metals below the threshold necessary to be considered hyperaccumulation (Van der Ent *et al.*, 2012; Krämer 2010).

Transfer of Metals into the Food Chain

Regardless of the reason for metal hyperaccumulation in plants, we must consider the implications of this trait in the environment around such plants. In particular, herbivores that feed on metal accumulating plants may transfer these metals into the food web, with potential for metals to bioaccumulate in higher trophic levels (Boyd, 2009; Cai *et al.*, 2009; Boyd, 2004; Peterson *et al.*, 2003; Wall and Boyd, 2002). Peterson *et al.* (2003) surveyed the arthropods of a serpentine outcrop in Portugal, where the Ni-hyperaccumulator *Alyssum pintodasilvae* is present, and found that Ni was being mobilized into the food chain. Similarly, Wall and Boyd (2002) found elevated levels of Ni in arthropods collected from a serpentine outcrop in the Red Hill formation of California, with the greatest concentrations of Ni being found in insects associated with the Ni-hyperaccumulator *S. polygaloides*. Outridge and Scheuhammer (1993) showed that vertebrates can be negatively affected if they consume >500 μg g^{-1} of Ni, suggesting that a diet rich in metal-laden insects may be harmful to birds and other animals.

Galeas *et al.* (2007) assessed arthropod abundance and diversity over two growing seasons in Se-enriched habitats in Colorado, comparing Se-hyperaccumulator species (including *S. pinnata*) with non-hyperaccumulator species. The Se-hyperaccumulators, with Se at 1,000 to 14,000 μg g^{-1} dry weight, harbored significantly fewer arthropods (approximately two-fold lower) and fewer arthropod species (approximately 1.5-fold lower) compared with non-hyperaccumulator species which contained Se at <30 μg g^{-1}. Arthropods collected on Se-hyperaccumulating plants contained three- to ten-fold higher concentrations of Se than arthropods found on non-hyperaccumulating species but less than ten-fold lower concentrations of Se than their hyperaccumulator host plants. Several arthropod species contained Se at >100 μg g^{-1}, indicating a relatively great tolerance to Se.

Some animals avoid metal-enriched tissue, lowering the chance that metals may enter the food chain (Boyd, 2007). Isopods fed with leaf litter from the hyperaccumulator *A. pintodasilvae* showed 83% mortality compared with isopods that were fed with leaf litter from non-hyperaccumulatoring plants. When given a choice in diet, the isopods preferred leaf litter from non-hyperaccumulating plants (Goncalves *et al.*, 2007). This finding supports the *elemental defense hypothesis* which suggests that certain animals may avoid eating metal-enriched plant tissue. Furthermore, a generalist diet in herbivorous and predacious insects may dilute any metals that the insect may have consumed from hyperaccumulating plants (Boyd, 2009). Further discussion of herbivory and other cross-kingdom interactions in metal-enriched environments can be found in Strauss and Boyd (2011).

CHALLENGES OF PHYTOREMEDIATION

Although phytoremediation using metal hyperaccumulators may be a low-cost, eco-friendly alternative to traditional cleanup methods, this technology is not without its limitations (Pilon-Smits, 2005). Firstly, hyperaccumulators must be able to grow on a contaminated site; the soil properties, concentration of metals, and climate cannot inhibit plant growth. Secondly, many naturally-occurring hyperaccumulators are small, with shallow root systems and minimal above-ground biomass, limiting the depth and amount of land that can be cleaned using a given plant. Thirdly, some metals may not be bioavailable in the soil and the soil may need to be chemically treated to make the metal bioavailable before introducing a hyperaccumulator for clean-up. Fourthly, most hyperaccumulators are specific for one metal only and many sites are contaminated with multiple metals. As discussed above, there is a risk that herbivores may transfer metals into the food chain. This risk, therefore, should be assessed on a site-by-site basis before implementing a phytoremediation effort (Neilson and Rajakaruna, 2012). Furthermore, the public may be concerned that non-native, fast-growing hyperaccumulators may escape from remediation sites and become invasive (Whiting *et al.*, 2004). As in a recent case in the town of O'Brien, Oregon, USA, *Alyssum murale* and *A. corsicum*—species from Mediterranean Europe—appear to have naturalized and become invasive in nearby serpentine outcrops, potentially threatening native plants: (http://www.oregon.gov/ODA/PLANT/WEEDS/edrr.shtml). Thus, it is vital to understand the biology and ecology of the plants to be used, in as much detail as possible, before undertaking field-based phytoremediation or phytomining operations using non-native species.

Given that many of the metal-hyperaccumulating species in the Brassicaceae are small-statured, there is an interest in genetically modifying hyperaccumulators to have greater biomass, deeper root systems, or an enhanced ability to uptake metals (Bhargava *et al.*, 2012; Cherian and Oliveira, 2005; Doty *et al.*, 2008; Pilon-Smits and Pilon, 2002; Rugh, 2004). A number of concerns have been raised regarding these designer hyperaccumulators (Ellstrand, 2001; Pilon-Smits and Freeman, 2006; Angle and Linacre, 2005; Eapen and D'Souza, 2005). As with naturally occurring hyperaccumulators, genetically modified (GM) hyperaccumulators may escape from their remediation sites and become invasive in other habitats (Pilon-Smits, 2005). Furthermore, pollen from GM and non-GM hyperaccumulators may be transferred off-site via wind or insects, landing on wild and agronomic relatives with the potential for metal-tolerant genes to become fixed in close relatives (Whiting *et al.*, 2004). Given that many hyperaccumulators and crop plants belong to the Brassicaceae (Warwick, 2012), this is a serious concern in need of further study (Neilson and Rajakaruna, 2012;

Whiting *et al.*, 2004). Species of the Brassicaceae that have been genetically modified to tolerate or hyperaccumulate heavy metals are listed in Table 3.

Table 3. Brassicaceae species that have been genetically modified to increase heavy metal accumulation

Species	Metals Hyperaccumulated	Reference
Arabidopsis spp.	As, Cd, Cu, Hg, Ni, Pb, Se, Zn	Li *et al.*, 2005 (As, Hg, Cd); Lee *et al.*, 2003 (Cd); Xu *et al.*, 2009 (Cu); Bizily *et al.*, 2003 (Hg); Pianelli *et al.*, 2005 (Ni); Song *et al.*, 2003 (Pb, Cd); Leduc *et al.*, (2004) (Se); Haydon and Cobbett, 2007 (Zn)
Brassica juncea	As, Cd, Pb, Se, Zn	Wangeline *et al.*, 2004 (As); Reisinger *et al.*, 2008 (Cd); Zhu *et al.*, 1999 (Cd); Bhuiyan *et al.*, 2011a,b (Pb); Gleba *et al.*, 1999 (Pb); Bañuelos *et al.* (Se), 2005; Bennett *et al.*, 2003 (Zn);
Brassica napus	As, Ni, Zn	Stearns *et al.*, 2005 (As); Brewer *et al.* (Ni), 1999; Nie *et al.*, 2002 (Zn)

One alternative to genetically modifying hyperaccumulators is to continue our search for undiscovered hyperaccumulators. There are potentially many more hyperaccumulators waiting to be discovered on both naturally occurring and anthropogenically created metal-enriched sites worldwide (Whiting *et al.*, 2004; Boyd *et al.*, 2009). Such sites are undergoing drastic changes due to ever-expanding development, deforestation, mining, exotic species invasions, and atmospheric deposition of various pollutants or previously limiting nutrients such as nitrogen (Williamson and Balkwill, 2006; Rajakaruna and Boyd, 2008; Harrison and Rajakaruna, 2011). Floristic surveys should be encouraged to document metal-tolerant and hyperaccumulating plants which may be at risk of being lost from these under-studied habitats worldwide.

Conclusion

The family Brassicaceae is extremely important for phytoremediation of heavy metals worldwide, both in the number of hyperaccumulating species found in this family and the knowledge they have supplied regarding metal tolerance at the molecular, cellular, and whole-plant level. The use of *Arabidopsis thaliana, A. halleri, Brassica juncea*, and *Noccaea caerulescens* as model species has revealed numerous physiological mechanisms contributing to metal uptake and has elucidated the genetics behind these mechanisms. Other species such as *Streptanthus polygaloides*, *Alyssum pintodasilvae*, and *Stanleya pinnata* have provided insight into the ecology of metal-tolerant species, helping assess the potential for transfer of metals into the food chain and to investigate the various hypotheses relating to the causes and consequences of metal hyperaccumulation. Despite extensive research over the last several decades, phytoremediation as a technique for cleaning metal contaminated sites needs further development. To improve this green technology, we must continue to research metal tolerance at the molecular, cellular, organismic, and ecosystem levels, address the risks these plants may pose to surrounding habitat, and actively search for new plant candidates for phytoremediation worldwide. Species from the Brassicaceae will no doubt continue to provide valuable insight on all aspects of metal-plant-ecosystem relationships.

ACKNOWLEDGEMENTS

We thank Dr. Robert S. Boyd, Dr. Alan J. M. Baker, and Tanner Harris for providing useful comments, Dr. Ihsan A. Al-Shehbaz for confirming the current nomenclature for taxa listed in Table 1, Dr. Roger D. Reeves for providing information on the metal hyperaccumulating Brassicaceae, Christopher Spagnoli for his assistance with Figure 2, and College of the Atlantic for providing funding via a Rothschild Faculty-Student Collaboration Grant.

REFERENCES

Adigüzel, N., Reeves, R. D., (2002). A new nickel-accumulating species of *Alyssum* (Cruciferae) from western Turkey. *Edinburgh J Bot.* 59, 215-219.

Al-Najar, H., Kaschl, A., Schulz, R., Römheld, V., (2005). Effects of thallium fractions in the soil and pollution origin in thallium uptake by hyperaccumulator plants: a key factor for assessment of phytoextraction. *Int J Phytoremediation*. 7, 55-67.

Anderson, C. W. N., Brooks, R. R., Chiarucci, A., LaCoste, C. J., Leblanc, M., Robinson, B. H., Simcock, R., Stewart, R. B., (1999). Phytomining for nickel, thallium and gold. *J. Geochem. Explor.* 67, 407-415.

Andres-Colas, N., Perea-Garcia, A., Puig, S., Penarrubia, L., (2010). Deregulated copper transport affects *Arabidopsis* development especially in the absence of environmental cycles. *Plant Physiol*. 153, 170-184.

Angle, J. S., Linacre, N. A., (2005). Metal phytoextraction – a survey of potential risks. *Int J Phytoremediation*. 7, 241-254.

Baker, A. J. M., McGrath, S. P., Reeves, R. D., Smith, J. A. C., (2000). Metal hyperaccumlator plants: a review of the ecology and physiology of a biological resource for phytoremediation of metal polluted soils. In: Terry, N., Bañuelos, G. (eds) *Phytoremediation of contaminated soil and water*. CRC Press, Boca Raton, pp 85-108.

Baker, A. J. M., Walker, P. L., (1989). Ecophysiology of metal uptake by tolerant plants. In: Shaw, A. J., (ed) *Heavy metal tolerance in plants - evolutionary aspects*. CRC Press, Boca Raton, pp 155-177.

Baker, A. J. M., (1981). Accumulators and excluders - strategies in the response of plants to heavy metals. *J Plant Nutr.* 3, 643-654.

Bani, A., Pavlova, D., Echevarria, G., Mullaj, A., Reeves, R. D., Morel, J. L., Sulçe, S., (2010). Nickel hyperaccumulaion by the species *Alyssum* and *Thlaspi* (Brassicaceae) from the ultramafic soils of the Balkans. *Bot Serbica*. 34, 3-14.

Bañuelos, G. S., (2001). The green technology of selenium phytoremediation. *Biofactors* 14, 255-260.

Bañuelos, G., Terry, N., Leduc, D. L., Pilon-Smits, E. A. H., Mackey, B., (2005). Field trial of transgenic mustard plants shows phytoremediation of selenium-contaminated sediment. *Environ Sci Technol.* 39, 1771-1777.

Barillas, J. R. V., Quinn, Q. F., Pilon-Smits, E. A. H., (2011). Selenium accumulation in plants – phytotechnological applications and ecological implications. *Int J Phytoremediation*. 13, 166-178.

Barzanti, R., Colzi, I., Arnetoli, M., Gallo, A., Pignattelli, S., Gabbrielli, R., Gonnelli, C., (2011). Cadmium phytoextraction potential of different *Alyssum* species. *J Hazard. Mater*. 196, 66-72.

Basic, N., Keller, C., Fontanillas, P., Vittoz, P., Besnard, G., Galland, N., (2006). Cadmium hyperaccumulation and reproductive traits in natural *Thlaspi caerulescens* populations. *Plant Biol*. 8, 64-72.

Becerra-Castro, C., Monterroso, C., Garcia-Leston, M., Prieto-Fernandez, A., Acea, M. J., Kidd, P. S., (2009). Rhizosphere microbial densities and trace metal tolerance of the nickel hyperaccumulator *Alyssum serpyllifolium* subsp. *lusitanicum*. *Int J Phytoremed.* 11, 525-541.

Becher, M., Talke, I. N., Krall, L., Kreamer, U., (2004). Cross-species microarray transcript profiling reveals high constitutive expression of metal homeostasis genes in shoots of the zinc hyperaccumulator *Arabidopsis halleri*. *Plant J.* 37, 251–268.

Bennett, L. E., Burkhead, J. L., Hale, K. L., Terry, N., Pilon, M., Pilon-Smits, E. A., (2003). Analysis of transgenic Indian mustard plants for phytoremediation of metal contaminated mine tailings. *J Environ Qual*. 32, 432-440.

Bert, V., Bonnin, I., Saumitou-Laprade, P., de Laguerie, P., Petit, D., (2002). Do *Arabidopsis halleri* from nonmetallicolous populations accumulate zinc and cadmium more effectively than those from metallicolous populations? *New Phytol*. 155, 47-57.

Bevan, M., Walsh, S., (2005). The *Arabidopsis* genome: A foundation for plant research. *Genome Res*. 15, 1632-1642.

Bhargava, A., Carmona, F. F., Bhargava, M., Srivastava S., (2012). Approaches for enhanced phytoextraction of heavy metals. *Jour. of Env. Manag*. 105, 103-120.

Bhuiyan, M. S. U., Min, S. R., Jeong, W. J., Sultana, S., Choi, K. S., Song, W. Y., Lee, Y., Lim, Y. P., Liu, J. R., (2011a). Overexpression of a yeast cadmium factor 1 (YCF1) enhances heavy metal tolerance and accumulation in Brassica juncea. *Plant Cell Tiss Org Cult*. 105, 85-91.

Bhuiyan, M. S. U., Min, S. R., Jeong, W. J., Sultana, S., Choi, K. S., Lee, Y., Liu, J. R., (2011b). Overexpression of AtATM3 in *Brassica juncea* confers enhanced heavy metal tolerance and accumulation. *Plant Cell Tiss Org Cult*. 107, 69-77.

Bidar, G., Pruvot, C., Garcon, G., Verdin, A., Shirali, P., Douay, F., (2009). Seasonal and annual variations of metal uptake, bioaccumulation, and toxicity in *Trifolium repens* and *Lolium perenne* growing in a heavy metal-contaminated field. *Environ Sci Pollut R*. 16, 42-53.

Bizily, S.P ., Kim, T., Kandasamy, M. K., Meagher, R. B., (2003). Subcellular targeting of methylHg lyase enhances its specific activity for organic Hg detoxification in plants. *Plant Physiol.* 131, 463-471.

Blaylock, M. J., Huang, J. W. (2000). Phytoextraction of metals. In: Raskin, I., Ensley, B. D., (eds) *Phytoremediation of toxic metals: using plants to clean up the environment*. Wiley Interscience New York, pp 53-70.

Boyd, R. S., (2004). Ecology of metal hyperaccumulation. *New Phytol.* 162, 563-567.

Boyd, R. S., (2007). The defense hypothesis of elemental hyperaccumulation: status, challenges and new directions. *Plant and Soil.* 293, 153-176.

Boyd, R. S., (2009), High-nickel insects and nickel hyperaccumulator plants: a review. *Insect Sci.*16, 19–31.

Boyd, R. S., (2010). Heavy metal pollutants and chemical ecology: exploring new frontiers. *J Chem Ecol.* 36, 46-58.

Boyd, R. S., Jaffré, T., (2001). Phytoenrichment of soil Ni concentration by *Sebertia acuminata* in New Caledonia and the concept of elemental allelopathy. *S Afr J Sci.* 97, 535-538.

Boyd, R. S., Martens, S. N., (1992). The raison d'etre for metal hyperaccumulation by plants. In: Baker, A. J. M., Proctor, J., Reeves, R. D. (eds), *The vegetation of ultramafic (serpentine) soils*, Intercept Limited, Andover, Hamshire, UK, pp 279-289.

Boyd, R. S., Martens, S. N., (1998). The significance of metal hyperaccumulation for biotic interactions. *Chemoecology.* 8, 1-7.

Boyd, R. S.,Wall, M. A., Santos, S. R., Davis, M. A. (2009). Variation of morphology and elemental concentrations in the California nickel hyperaccumulator *Streptanthus polygaloides* (Brassicaceae). *Northeast Nat.* 16, 21-38.

Boyd, R. S., Kruckeberg, A. R., Rajakaruna, N., (2009). Biology of ultramafic rocks and soils: research goals for the future. *Northeast Nat.* 16, 422-440.

Brewer, E. P., Saunders, J. A., Angle, J. S., Chaney, R. L., McIntosh, M. S., (1999). Somatic hybridization between the Zn accumulator *Thlaspi caerulescens* and *Brassica napus. Theor Appl Genet.* 99, 761-771.

Broadhurst, C. L., Chaney, R. L., Angle, J. S., Maugel, T. K., Erbe, E .F., Murphy, C. A., (2004). b. Simultaneous hyperaccumulation of nickel, manganese, and calcium in *Alyssum* leaf trichomes. *Environ. Sci. Technol.* 38, 5797–5802.

Broadley, M. R., White, P. J., Hammond, J. P., Zelko, I., Alexander Lux., (2006). Zinc in plants. *New Phytol.* 173, 677–702.

Brooks R. R., Radford, C. C., (1978). Nickel accumulation by European species of the genus *Alyssum*. *Proc Roy Soc London*. B200, 217-224.

Brooks, R. R., Morrison, R. S., Reeves, R. D ., Dudley, T. R. and Akman, Y., (1979). Hyperaccumulation of nickel by *Alyssum* Linnaeus (Cruciferae). *Proc. Roy. Soc. London.* B203, 387-403.

Caesalpino, A., (1583). De Plantis Libri, Vol. 16, Florentiae, p. 369.

Cai, Q., Long, M. L., Zhu, M., Zhou, Q. Z., Zhang, L., Liu, J., (2009). Food chain transfer of cadmium and lead to cattle in a lead-zinc smelter in Guizhou, China. *Environ Pollut.* 157, 3078-3082.

Cecchi, L., Gabbreilli, R., Arnetoli, M., Gonnelli, C., Hasko, A., Selvi, F., (2010). Evolutionary lineages of nickel hyperaccumulation and systrmatics in European Alysseae (Brassicaceae): evidence from nrDNA sequence data. *Ann Bot.* 106, 751-767.

Chaffai, R., Koyama, H., (2011). Heavy metal tolerance in *Arabidopsis thaliana. Adv Bot Res.* 60, 1-49.

Chao, Y. Fu, D., (2004). Kinetic study of the antiport mechanism of an *Escherichia coli* zinc transporter, ZitB. *J Biol Chem.* 279, 12043–12050.

Cherian, S., Oliveira, M.M., (2005). Transgenic plants in phytoremediation: Recent advances and new possibilities. *Environ. Sci. Technol.* 39, 9377-9390.

Cho, M., Chardonnens, A. N., Dietz. K-J., (2003). Differential heavy metal tolerance of *Arabidopsis halleri* and *Arabidopsis thaliana*: a leaf slice test. *New Phytol.* 158, 287–293.

Cobbett, C. S., (2000). Phytochelatins and their role in heavy metal detoxification. *Plant Physiol.* 123. 825–33.

Cobbett, C., Goldsbrough, P., (2002). Phytochelatins and metallothioneins: roles in heavy metal detoxification and homeostasis. *Annu Rev Plant Biol.* 53, 159-182.

Cole, M. M., (1973). Geobotanical and biogeochemical investigations in the sclerophyllous woodland and shrub associations of the Eastern Goldfields area of Western Australia, with particular reference to the role of *Hybanthus floribundus* (Lindl.) F. Muell. as a nickel indicator and accumulator plant. *J Appl Ecol.* 10, 269-320.

Cosio, C., Martinoia, E., Keller, C., (2004). Hyperaccumulation of cadmium and zinc in *Thlaspi caerulescens* and *Arabidopsis halleri* at the leaf and cellular level. *Plant Physiol.* 134, 716-725.

Courbot, M., Willems, G., Motte, P., Arvidsson, S., Roosens, N., Saumitou-Laprade, P., Verbruggen, N., (2007). A major quantitative trait locus for cadmium tolerance in *Arabidopsis halleri* colocalizes with HMA4, a gene encoding a heavy metal ATPase. *Plant Physiol.* 144, 1052-1065.

Das, S., Sen, M., Saha, C., Chakraborty, D., Das, A., Banerjee, M., Seal, A., (2011). Isolation and expression analysis of partial sequences of heavy metal transporters from Brassica juncea by coupling high throughput cloning with a molecular fingerprinting technique. *Planta.* 234, 139-156.

DiDonato, R. J., Roberts, L. A., Sanderson, T., Eisley, R. B., Walker, E. L., (2004). Arabidopsis Yellow Stripe-Like2 (YSL2): a metal-regulated gene encoding a plasma membrane transporter of nicotianamine-metal complexes. *Plant J.* 39, 403-414.

Doksopulo, E. P. 1961. Nickel in rocks, soils, water and plants adjacent to the talc deposits of the Chorchanskaya group. Izdatel vo Tbiliskovo Universitet, Tbilisi.

Doty, S. L., (2008). Enhancing phytoremediation through the use of transgenics and endophytes. *New Phytol.* 179, 318–33.

Eapen, S., D'Souza, S. F., (2005). Prospects of genetic engineering of plants for phytoremediation of toxic metals. *Biotechnol Adv.* 23, 97-114.

El Mehdawi, A. F., Pilon-Smits, E. A., (2012). Ecological aspects of plan selenium hyperaccumulation. *Plant Biol (Stuttg).* 14, 1-10.

Ellstrand, N. C. (2001). When transgenes wander, should we worry? *Plant Physiol.* 125, 1543-1545.

Ensley, B. D., (2000). Phytoremediation for toxic metals – using plants to clean-up the environment. In: Raskin, I., Ensley, B. D. (eds), *Rational(e?) for use of phytoremediation.* John Wiley & Sons, Inc., pp 3-13.

Epstein, E., Bloom, A.J., (2004). *Mineral nutrition of plants: principles and perspectives.* 2nd Edn. Sunderland: Sinauer Associates.

Ernst, W. H. O., (1966). Ökologisch-soziologische Untersuchungen an Schwermetallpflanzen-gesellschaften Südfrankreichs und des östlichen Harzvorlandes. *Flora, Jena Bl* 56, 301-318.

Ernst, W. H. O., (1968). Das Violetum calaminariae westfalicum, eine Schwermetallpflanzengesellschaft bei Blankenrode in Westfalen. *Mitteilungen der Floristisch-soziologischen Arbeitsgemeinschaft* 13, 263-268.

Ernst, W. H. O., (1974). Schwermetallvegetation der Erde. Fischer, Stuttgart.

Escarré, J., Lefèbvre, C., Gruber, W., Leblanc, M., Lepart, J., Rivière, Y., Delay, B., (2000). Zinc and cadmium accumulation by *Thlaspi caerulescens* from metalliferous and non-metalliferous sites in the Mediterranean area: implications for phytoremediation. *New Phytol*. 145, 429-437.

Farago, M. E., Cole, M. M., (1988). Nickel and plants. In: Sigel, H., Sigel, A., (eds), *Metal ions in biological systems*, vol. 23, Nickel and its role in biology, Marcel Dekker, New York, New York, pp 47-90.

Freeman, J. L., Tamaoki, M., Stushnoff, C., Quinn, C.F., Cappa, J. J., Devonshire, J., Fakra, S. C., Marcus, M. A., McGrath, S. P., Hoewyk, D. V., Pilon-Smits, E. A. H., (2010). Molecular mechanisms of selenium tolerance and hyperaccumulation in *Stanleya pinnata*. *Plant Physiol*. 153, 1630-1652.

Friesen, M. L., Von Wettberg, E. J., (2010). Adapting genomics to study the evolution and ecology of agricultural systems. *Curr. Opin. Pl. Biol*. 13, 119–125.

Galeas, M. L., Zhang, L-H., Freeman, J. L., Wegner, M., Pilon-Smits, E. A. H., (2007). Seasonal fluctuations of selenium and sulfur accumulation in selenium hyperaccumulators and related nonaccumulators. *New Phytol*. 173, 517-525.

Gendre, D., Czernic, P., Conéjéro, G., Pianelli, K., Briat, J.-F., Lebrun, M., Mari, S., (2006). TcYSL3, a member of the YSL gene family from the hyperaccumulator *Thlaspi caerulescens,* encodes a nicotianamine-Ni/Fe transporter. *Plant J*. 49, 1-15.

Ghaderian, S. M., Mohtadi, A., Rahiminejad, M. R. and Baker, A. J. M., (2007a). Nickel and other metal uptake and accumulation by species of *Alyssum* from the ultramafics of Iran. *Environ. Pollut.* 145,293-298.

Ghaderian, S. M., Mohtadi, A., Rahiminejad, R., Reeves, R. D., Baker, A. J. M., (2007b). Hyperaccumulation of nickel by two *Alyssum* species from the serpentine soils of Iran. *Plant Soil*. 293, 91-97.

Ghasemi, R., Ghaderian, S. M., Kramer, U., (2009). Accumulation of nickel in trichomes of a nickel hyperaccumulator plant, *Alyssum inflatum*. *Northeast Nat*. 16, 81-92.

Ghosh, M., Singh, S. P., (2005). A review on phytoremediation of heavy metals and utilization of its byproducts. *App Ecol Environ Res*. 3, 1-18.

Gleba, D., Borisjuk, N. V., Borisjuk, L. G., Kneer, R., Poulev, A., Skarzhinskaya, M., Dushenkov, S., Logendra, S., Gleba, Y. Y., Raskin, I., (1999). Use of plant roots for phytoremediation and molecular farming. Proc Natl Acad Sci. (USA) 96, 5973-5977.

Goncalves, M., Goncalves, S., Portugal, A., Silva, S., Sousa, J., Freitas, H., (2007). Effects of nickel hyperaccumulation in *Alyssum pintodasilvae* on model arthropods representative of two trophic levels. *Plant Soil*. 293, 177-188.

Gravot, A., Lieutaud, A., Verret, F., Auroy, P., Vavasseur, A., Richaud, P., (2004). AtHMA3, a plant P_{1B}-ATPase, functions as a Cd/Pb transporter in yeast. *FEBS Lett*. 561, 22-28.

Ha, S. B., Smith, A. P., Howden, R., Dietrich, W. M, Bugg, S., O'Connell, M. J., Goldsbrough, P. B., Cobbett, C. S., (1999). Phytochelatin synthase genes from *Arabidopsis* and the yeast. *Schizosaccharomyces pombe*. *Plant Cell.* 11, 1153–1164.

Hall, J. L., Williams, L. E., (2003). Transition metal transporters in plants. *J. Exp. Bot*. 54, 2601–2613.

Hanikenne, M., Talke, I. N., Haydon, M. J., Lanz, C., Nolte, A., Motte, P., Kroymann, J., Weigel, D., Krämer, U., (2008). Evolution of metal hyperaccumulation required cis-regulatory changes and triplication of HMA4. *Nature*. 453, 391-395.

Harrison, S. P., Rajakaruna, N., (2011). What have we learned from serpentine about evolution, ecology, and other sciences? In: Harrison, S. P., Rajakaruna, N., (eds) *Serpentine: evolution and ecology in a model system*, University of California Press, Berkeley, CA, USA, pp 417-427.

Haydon, M. J., Cobbett, C. S., (2007). A novel major facilitator superfamily protein at the tonoplast influences Zn tolerance and accumulation in *Arabidopsis*. *Plant Physiol.* 143, 1705-1719.

Heiss, S., Wachter, A., Bogs, J., Cobbett, C., Rausch, T., (2003). Phytochelatin synthase (PCS) protein is induced in *Brassica juncea* leaves after prolonged Cd exposure. *J Exp Bot*. 54, 1833-1839.

Hladun, K. R., Parker, D. R., Trumble, J. T., 2011. Selenium accumulation in the floral tissues of two Brassicaceae species and its impact on floral traits and plant performance. *Environ Exp Bot*. 74, 90-97

Hossain, M. A., Piyatida, P., Teixeria da Silva, J. A., Fujita, M., (2012). Molecular mechanism of heavy metal toxicity and tolerance in plants: Central role of glutathione in detoxification of reactive oxygen species and methylglyoxal and in heavy metal chelation. *J. Bot*. pp 37.

Hughes, R., Bachmann, K., Smirnoff, N., Macnair, M. R., (2001). The role of drought tolerance in serpentine tolerance in the *Mimulus guttatus* Fischer ex DC. complex. *S Afr J Sci* 97, 581–586.

Jabeen, R., Ahmad, A., Iqbal, M., (2009). Phytoremediation of heavy metals: physiological and molecular mechanisms. *Bot. Rev*. 75, 339-364.

Jhee, E. M., Boyd, R. S., Eubanks, M. D., (2005). Nickel hyperaccumulation as an elemental defense of *Streptanthus polygaloides* (Brassicaceae): Influence of herbivore feeding mode. *New Phytol.* 168:331–344.

Kabata-Pendias, A., Pendias, H., (2001). Trace Elements in Soils and Plants. 3rd ed, CRC Press, Boca Raton, FL.

Kawachi, M., Kobae, Y., Mimura, T., Maeshima, M., (2008). Deletion of a histidine-rich loop of AtMTP1, a vacuolar Zn^{2+}/H^{+} antiporter of *Arabidopsis thaliana*, stimulates the transport activity. *J Biol Chem*. 283, 8374-8383.

Kerkeb, L., Mukherjee, I., Chatterjee, I., Lahner, B., Salt, D. E., Connolly, E. L., (2008). Iron-induced turnover of the Arabidopsis Iron-Regulated Transporter1 metal transporter requires lysine residues. *Plant Physiol*. 146, 1964-1973

Kim, D., Gustin, J. L., Lahner, B., Persans, M. W., Baek, D., Yun, D. J., Salt, D. E., (2004). The plant CDF family member TgMTP1 from the Ni/Zn hyperaccumulator *Thlaspi goesingense* acts to enhance efflux of Zn at the plasma membrane when expressed in *Saccharomyces cerevisiae*. *Plant J*. 39, 237-251.

Kobayashi, Y., Kuroda, K., Kimura, K., Southron-Francis, J. L., Furuzawa, A.,Kimura, K., Iuchi, S., Kobayashi, M., Taylor, G. J., Koyama, H., (2008). Amino acid polymorphisms in strictly conserved domains of a P-type ATPase HMA5 are involved in the mechanism of copper tolerance variation in *Arabidopsis*. *Plant Physiol*. 148, 969–980.

Krämer, U., (2005). Phytoremediation: novel approaches to cleaning up polluted soils. *Curr Opin Biotechnol*. 16, 133-141.

Krämer, U., (2010). Metal hyperaccumulation in plants. *Annu. Rev. Plant Biol.* 61, 517–534.

Kruckeberg A. R., Reeves, R. D., (1995). Nickel accumulation by serpentine species of *Streptanthus* (Brassicaceae): field and greenhouse studies. *Madroño*. 42, 458-469.

Kubota, H., Takenaka, C., (2003). *Arabis gemmifera* is a hyperaccumulator of Cd and Zn. *Int J Phytoremed.* 5, 197-201.

Kumar, P. B. A. N., Dushenkov, V., Motto, H., Raskin, I., (1995). Phytoextraction: the use of plants to remove heavy metals from soils. *Environ. Sci. Technol.* 29, 1232-1238.

Küpper, H., Kochian, L. V., (2010). Transcriptional regulation of metal transport genes and mineral nutrition during acclimatization to cadmium and zinc in the Cd/Zn hyperaccumulator, *Thlaspi caerulescens* (Ganges population). *New Phytol.* 185, 114-129.

Lasat, M. M., Kochian, L. V., (2000). Physiology of Zn hyperaccumulation in *Thlaspi caerulescens*. In: Terry, N., Bañuelos, G., (eds) Phytoremediation of contaminated soil and water. CRC Press, Boca Raton, pp 167–177.

Leblanc, M., Petit, D., Deram, A., Robinson, B., Brooks, R. R., (1999). The phytomining and environmental significance of hyperaccumulation of thallium by *Iberis intermedia* from southern France. *Econ Geol.* 94, 109-113.

LeDuc, D. L., Tarun, A. S., Montes-Bayon, M., Meija, J., Malit, M. F., Wu, C. P., AbdelSamie, M., Chiang, C-Y., Tagmount, A., deSouza, M., Neuhierl, B., Bock, A., Caruso, J., Terry, N., (2004). Overexpression of selenocysteine methyltransferase in *Arabidopsis* and Indian mustard increases selenium tolerance and accumulation. *Plant Physiol.* 135, 377–383.

Lee, S., Moon, J. S., Ko, T.S., Petros, D., Goldsbrough, P. B. and Korban, S. S. (2003). Overexpression of *Arabidopsis* phytochelatin synthase paradoxically leads to hypersensitivity to cadmium stress. *Plant Physiol.* 131, 656–663.

Li, Y., Dhankher, O. P., Carreira, L., Balish, R. S., Meagher, R. B., (2005). Arsenic and Hg tolerance and Cd sensitivity in *Arabidopsis* plants expressing bacterial gammaglutamylcysteine synthetase. *Environ Toxicol Chem.* 24, 1376-1386.

Lim, J. M., Salido, A. L., Butcher, D. J., (2004). Phytoremediation of lead using Indian mustard (*Brassica juncea*) with EDTA and electrodics. *Microchem J.* 76, 3-9.

Lombi, E., Zhao, F. J., Dunham, S. J., and McGrath, S. P., (2000). Cadmium accumulation in populations of *Thlaspi caerulescens* and *Thlaspi goesingense*. *New Phytol.* 145, 11-20.

Maestri, E., Marmiroli, M., Visioli G., Marmiroli, N., (2010). Metal tolerance and hyperaccumulation: Costs and trade-offs between traits and environment. *Env. and Exp. Botany.* 68, 1-13.

Martens, S. N., Boyd, R. S., (2002). The defensive role of Ni hyperaccumulation by plants: a field experiment. *Am J Bot.* 89, 998-1003.

Memon, A. R., Schroeder, P., (2009). Implications of metal accumulation mechanisms to phytoremediation. *Environ. Sci. Pollut. Res.* 16, 162-175.

Mench, M., Schwitzguebel, J. P., Schroeder, P., Bert, V., Gawronski, S., Gupta, S., (2009). Assessment of successful experiments and limitations of phytotechnologies: contaminant uptake, detoxification and sequestration, and consequences for food safety. *Environ Sci Pollut Res.* 16, 876-900.

Mendoza-Cózatl, D. G., Jobe, T., Hauser, F., Schroeder, J. I., (2011). Long-distance transport, vacuolar sequestration, tolerance, and transcriptional responses induced by cadmium and arsenic. *Curr Opin Plant Biol.* 14, 554-562.

Meyer C.-L. and Verbruggen N., (2012). Use of the model species *Arabidopsis halleri* towards phytoextraction of cadmium polluted soils. *N Biotechnol* 30, 9-14.

Milner, M. J., Kochian, L. V., (2008). Investigating heavy-metal hyperaccumulation using *Thlaspi caerulescens* as a model system. *Ann. Bot.* 102, 3-13.

Minguzzi, C. and Vergnano, O., (1948). Il contenuto di nichel nelle ceneri di *Alyssum bertolonii* Desv.. *Atti della Societa Toscana di Scienze Naturali, Memorie Serie A* 55, 49-77.

Mizuno, T., Usui, K., Horie, K., Nosaka, S., Mizuno, N., Obata, H., (2005). Cloning of three ZIP/NRAMP transporter genes from a Ni hyperaccumulator plant *Thlaspi japonicum* and their Ni 2fl-transport abilities. Plant Physiol. Biochem. 43, 793-801.

Morel, M., Crouzet, J., Gravot, A., Auroy, P., Leonhardt, N., Vavasseur, A., Richaud, P., (2008). AtHMA3, a P_{1B}-ATPase allowing Cd/Zn/Co/Pb vacuolar storage in Arabidopsis. *Plant Physiol*. 149, 894-904.

Moxon, A. L., Olson, O. E., Seawright, W. V., (1950). Selenium in rocks, soils and plants. South Dakota Agricultural Experimental Station, Revised Technical Bulletin 2, 1-94.

Neilson, S., Rajakaruna, N., (2012). Roles of rhizospheric processes and plant physiology in phytoremediation of contaminated sites using oilseed Brassicas. In: Anjum N. A., Ahmad I., Pereira M. E., Duarte A. C., Umar S., Khan N. A. (eds) *The plant family Brassicaceae: contribution towards phytoremediation*. Environmental Pollution Book Series, Vol. 21, Springer, Dordrecht, The Netherlands, pp 313-330.

Nie, L., Shah, S., Rashid, A., Burd, G. I., Dixon, D. G., Glick, B. R., (2002). Phytoremediation of arsenate contaminated soil by transgenic canola and the plant growth promoting bacterium *Enterobacter cloacae* CAL2. *Plant Physiol Biochem*. 40, 355-361.

O'Dell, R. E., Rajakaruna, N., (2011). Intraspecific variation, adaptation, and evolution. In: Harrison, S. H., Rajakaruna, N., (eds) *Serpentine: the evolution and ecology of a model system*, Univ. of California Press, Berkeley, pp 97-137.

Outridge, P. M., Scheuhammer, A. M., (1993). Bioaccumulation and toxicology of chromium: implications for wildlife. *Rev. Environ. Contam. Toxicol*. 130, 31-77.

Patrick, L., (2006). Lead toxicity, a review of the literature. Part 1: Exposure, evaluation, and treatment. *Alt. Med. Rev.: a J. Clin. Therap.* 11, 2-22.

Peterson, L. R., Trivett, V., Baker, A. J. M., Aguiar, C., Pollard, A. J., (2003). Spread of metals through an invertebrate food chain as influenced by a plant that hyperaccumulates nickel. *Chemoecology*. 13, 103-108.

Pianelli, K., Mari, S., Marquès, L., Lebrun, M., Czernic, P., (2005). Nicotianamine over-accumulation confers resistance to nickel in *Arabidopsis thaliana*. *Transgenic Res.* 14, 739-748.

Pilon-Smits, E., (2005). Phytoremediation. *Annu. Rev. Plant Biol*. 56, 15-39.

Pilon-Smits, E. A. H., Freeman, J. L., (2006). Environmental cleanup using plants: biotechnological advances and ecological considerations. *Front Ecol Environ*. 4, 203–210.

Pilon-Smits, E., Pilon, M., (2002). Phytoremediation of metals using transgenic plants. *Critical Reviews in Plant Sciences*. 21, 439-456.

Plaza, S., Tearall, K. L., Zhao, F. J., Buchner, P., McGrath, S. P., Hawkesford, M. J., (2007). Expression and functional analysis of metal transporter genes in two contrasting ecotypes of the hyperaccumulator *Thlaspi caerulescens*. *J Exp Bot*. 58, 1717-1728.

Pope, N., Fong, M., Boyd, R., Rajakaruna, N. (2013). The role of elevation and soil chemistry in the distribution and ion accumulation of floral morphs of *Streptanthus polygaloides* Gray (Brassicaceae), a Californian nickel hyperaccumulator. *Plant Ecol Divers., DOI:10.1080/17550874.2013.783141.*

Puig, S., Andres-Colas, N., Garcia-Molina, A., and Penarrubia, L. (2007). Copper and iron homeostasis in *Arabidopsis*: responses to metal deficiencies, interactions and biotechnological applications. *Plant Cell Environ*. 30, 271-290.

Puschenreiter, M., Wieczorek, S., Horak, O., Wenzel, W. W., (2003). Chemical changes in the rhizosphere of metal hyperaccumulator excluder *Thalspi* species. *J Plant Nutr Soil Sci*. 168, 579-584.

Rascio, W., (1977). Metal accumulation by some plants growing on Zn mine deposits. *Oikos* 29, 250-253.

Rascio, N., Navari-Izzo, F., (2011). Heavy metal accumulating plants: how and why do they do it? And what makes them so interesting? *Plant Sci*. 180, 169–181.

Rajakaruna, N., Boyd, R. S., (2008). Edaphic Factor. In: Jorgensen, S. E., Fath, B. D. (eds), *General ecology*, Volume 2 of Encyclopedia of Ecology, Elsevier Science, Amsterdam, The Netherlands, pp 1201-1207.

Rajakaruna N., Tompkins, K. M., Pavicevic, P. G., (2006). Phytoremediation: an affordable green technology for the clean-up of metal contaminated sites in Sri Lanka. *Ceylon J Sci*. 35, 25–39.

Reeves, R. D., (1988). Nickel and zinc accumulation by species of *Thlaspi* L., *Cochlearia*, L. and other genera of the Brassicaceae. *Taxon*. 37, 309-318.

Reeves, R. D., Adigüzel, N., (2008). The nickel hyperaccumulating plants of the serpentines of Turkey and adjacent areas: a review with new data. *Turk J Biol*. 32, 143-153.

Reeves, R. D., Adigüzel, N., Baker, A. J. M., (2009). Nickel hyperaccumulation in *Bornmuellera kiyakii* and associated plants of the Brassicaceae from Kızıldaäÿ Derebucak (Konya), Turkey. *Turk J Bot*. 33, 33-40.

Reeves, R. D., Kruckeberg, A. R., Adıgüzel, N. Krämer, U., (2001). Studies on the flora ofserpentine and other metalliferous areas of western Turkey. *S Afr J Sci*. 97, 513-517.

Reeves, R. D., Baker, A. J. M., (2000). Metal-accumulating plants. In: Raskin, I., Ensley, B. D. (eds) *Phytoremediation of toxic metals: using plants to clean up the environment*. Wiley, New York, pp 193–229.

Reeves, R. D., Baker, A. J. M., Kelepertsis, A. (1997). The distribution and biogeochemistry of some serpentine plants of Greece. In: Jaffré, T., Reeves, R.D., Becquer, T. (eds). *Écologie des milieux sur roches ultramafiques et sur sols métallifères*, ORSTOM, Nouméa, Documents Scientifiques et Techniques No. III/2. pp 205-207.

Reeves, R. D., Baker, A. J. M., (1984). Studies on metal uptake by plants from serpentine and non-serpentine populations of *Thlaspi goesingense* Halacsy (Cruciferae). *New Phytol*. 98,191-204.

Reeves, R. D., Brooks, R. R., (1983). European species of *Thlaspi* L. (Cruciferae) as indicators of nickel and zinc. *J Geochem Explor*. 18, 275-283.

Reeves, R. D., Brooks, R. R., Dudley, T. R. (1983). Uptake of nickel by species of *Alyssum*, *Bommuellera* and other genera of Old World tribus Alysseae. *Taxon*. 32, 184-192.

Reeves, R. D., Brooks, R. R., Macfarlane, R. M., (1981). Nickel uptake by Californian *Streptanthus* and *Caulanthus* with particular reference to the hyperaccumulator *S. polygaloides* Gray (Brassicaceae). *Am J Bot*. 68, 708–712.

Reeves, R. D., Brooks, R. R., Press, J. R., (1980). Nickel accumulation by species of *Peltaria* Jacq. (Cruciferae). *Taxon*. 29, 629-633.

Reisinger, S., Schiavon, M., Terry, N., Pilon-Smits, E. A., (2008). Heavy metal tolerance and accumulation in Indian mustard (*Brassica juncea* L.) expressing bacterial gamma-glutamylcysteine synthetase or glutathione synthetase. *Int J Phytoremed*. 10, 440-454.

Richau, K. H., Kozhevnikova, A. D., Seregin, I. V., Vooijs, R., Koevoets, P. M., Smith, J. A., Ivanov, V. B., Schat, H., (2009). Chelation by histidine inhibits the vacuolar sequestration of nickel in roots of the hyperaccumulator *Thlaspi caerulescens*. *New Phytol*. 183, 106-116.

Roosens, N. H.. Willems, G., Saumitou-Laprade, P., (2008a). Using *Arabidopsis* to explore zinc tolerance and hyperaccumulation. *Trends Plant Sci.* 13, 208-215.

Roosens, N. H. C. J., Glenda, W., Cécile, G., Adeline, C., Pierre, S. L., (2008b). The use of comparative genome analysis and synthetic relationships allows extrapolating the position of Zn tolerance QTL regions from *Arabidopsis halleri* into *Arabidopsis thaliana*. *Plant Soil*. 306, 105-116.

Rosenfeld, I., Beath, O. A., (1964). *Selenium - geobotany, biochemistry, toxicity and nutrition.* Academic Press, New York.

Rugh, C. L. (2004). Genetically engineered phytoremediation: one man's trash is another man's transgene. *Trends Biotechnol*. 22, 496–498.

Sachs, J., (1865). *Handbuch der Experimental-Physiologie der Pflanzen.* Wilhelm Engelmann Verlag, Leipzig, Germany, pp 153-154..

Salt, D. E., Kato, N., Kramer, U., Smith, R. D., Raskin, I., (2000). The role of root exudates in nickel hyperaccumulation and tolerance in accumulator and nonaccumulator species of Thlaspi. In: Terry, N., Bañuelos, G., (eds) *Phytoremediation of contaminated soil and water*. CRC Press, Boca Raton, pp. 196–207.

Salt, D. E., Rauser, W. E., (1995). MgATP-dependent transport of phytochelatins across the tonoplast of oat roots. *Plant Physiol*. 107, 1293-1301.

Sancenon, V., Puig, S., Mateu-Andres, I., Dorcey, E., Thiele, D.J., Penarrubia, L., (2004). The Arabidopsis copper transporter COPT1 functions in root elongation and pollen development. *J Biol Chem*. 279, 15348-15355.

Schikora, A., Thimm, O., Linke, B., Buckhout, T. J., Mu̇ller, M., Schmidt, W., (2006). Expression, localization, and regulation of the iron transporter LeIRT1 in tomato roots. *Plant Soil*. 284, 101-108.

Song, W. Y., Sohn, E. J., Martinoia, E., Lee, Y. J., Yang, Y. Y., Jasinski, M., Forestier, C., Hwang, I., Lee, Y., (2003). Engineering tolerance and accumulation of Pb and Cd in transgenic plants. *Nat. Biotech*. 21, 914-919.

Strauss, S. Y., Boyd, R. S., (2011). Herbivory and other cross-kingdom interactions on harsh soils. In: Harrison, S. H., Rajakaruna, N., *Serpentine: the evolution and ecology of a model system* (eds) Univ. of California Press, Berkeley, pp 181-199.

Szczyglowska, M., Piekarska, A., Konieczka, P., Namiesnik, J. (2011). Use of *Brassica* Plants in the Phytoremediation and Biofumigation Processes. *Int J Mol Sci.* 12, 7760-7771.

Talke, I., Hanikenne, M., Krämer, U., (2006). Zn dependent global transcriptional control, transcriptional de-regulation and higher gene copy number for genes in metal homeostasis of the hyperaccumulator *Arabidopsis halleri*. *Plant Physiol*. 142, 148-167.

Tang, Y.-T., Qiu, R.-L., Zeng, X.-W., Ying, R.-R., Yu, F.-M., Zhou, X.-Y. (2009). Lead, zinc cadmium accumulation and growth simulation in *Arabis paniculata*. *Franch Env Exp. Bot*. 66, 126-134.

Taylor, S. I., Macnair, M., (2006). Within and between population variation for zinc and nickel accumulation in two species of *Thlaspi* (Brassicaceae). *New Phytol.* 169, 505-514.

Terry, N., Zayed, A. M., de Souza, M. P., Tarun, A. S. (2000). Selenium in higher plants, *Annu. Rev. Plant Physiol. Plant Mol. Biol.* 51, 401–432.

Thapa, G., Sadhukhan A., Panda, S. K. Sahoo, L., (2012). Molecular mechanistic model of plant heavy metal tolerance. *Biometals*. 25. 489–505.

Thomine, S., Lelièvre, F., Debarbieux, E., Schroeder, J. I., Barbier, Brygoo, H. (2003). AtNRAMP3, a multispecific vacuolar metal transporter involved in plant responses to iron deficiency. *The Plant J.* 34, 685–695.

Tsao, D. T., (2003) Phytoremediation. In: Scheper, T., Tsao, D.T. (eds), *Advances in biochemical engineering biotechnology*, Vol 78, p 206.

Vamerali, T., Bandiera, M., Mosca, G., (2010). Field crops for phytoremediation of metal-contaiminated land. A review. *Environ Chem Lett.* 8, 1-17.

Van der Ent, A., Baker, A. J. M., Reeves, R. D., Pollard, A. J., Schat, H., (2012). Hyperaccumulators of metal and metalloid trace elements: Facts and fiction. *Plant Soil*. 1-16

Van de Mortel, J. E., Villanueva, L. A., Schat, H., Kwekkeboom, J., Coughlan, S., Moerland, P. D., Ver Loren van Themaat, E., Koornneef, M., Aarts, M. G. M., (2006). Large expression differences in genes for iron and Zn homeostasis, stress response, and lignin biosynthesis distinguish roots of *Arabidopsis thaliana* and the related metal hyperaccumulator *Thlaspi caerulescens*. *Plant Physiol.* 142, 1127-1147.

Vatamaniuk O. K., Mari, S., Lu, Y. P., Rea, P. A., (1991). AtPCS1, a phytochelatin synthase from *Arabidopsis*: isolation and *in vitro* reconstitution. *Proc. of the Nat. Academy of Sci.* 96, 7110–7115.

Verbruggen, N., Hermans, C., Schat, H., (2009). Molecular mechanisms of metal hyperaccumulation in plants. *New Phytol.* 181, 759-776.

Vergnano Gambi, O., Gabbrielli, R., (1979). Ecophysiological and geochemical aspects of nickel, chromium and cobalt accumulation in the vegetation of some Italian ophiolitic outcrops. *Ofioliti*. 4, 199-208.

Verret, F., Gravot, A., Auroy, P., Leohardt, N., David, P., Nussaume, L., Vavasseur, A., Richaud, P., (2004). Overexpression of AtHMA4 enhances root-to-shoot translocation of zinc and cadmium and plant metal tolerance. *FEBS Lett.* 576, 306-312.

Wall, M. A., Boyd, R. S. (2002). Nickel accumulation in serpentine arthropods from the Red Hills, California. *Pan-Pac Entomol.* 78, 168-176.

Wangeline, A. L., Burkhead, J. L., Hale, K. L., Lindblom, S. D., Terry, N., Pilon, M., Pilon-Smits, E. A. H., (2004). Overexpression of ATP sulfurylase in Indian mustard: effects on tolerance and accumulation of twelve metals. *J Environ Qual.* 33, 54-60.

Warwick, S. I., (2011) Brassicaceae in Agriculture. In: Schmidt, R., Bancroft, I., (eds), *Genetics and genomics of the brassicaceae.* Plant genetics and genomics: Crop models, Vol 9, Springer, pp 33-65.

Weber, M., Harada, E., Vess, C., Roepenack-Lahaye, E.V., Clemens, S., (2004). Comparative microarray analysis of *Arabidopsis thaliana* and *Arabidopsis halleri* roots identifies nicotianamine synthase, a ZIP transporter and other genes as potential metal hyperaccumulation factors. *Plant J.* 37, 269–281.

Weigel, D., Mott, R., (2009). The 1001 genomes project for *Arabidopsis thaliana*. *Gen. Bio.* 10, 107.

Whiting, S. N., Reeves, R. D., Richards, D., Johnson, M. S., Cooke, J. A., Malaisee, F., Paton, A., Smith, J. A. C., Angle, J. S., Chaney, R. L., Ginoocchio, R., Jaffré, T., Johns, R., McIntyre, T., Purvis, O. W., Salt, D. E., Schat, H., Zhao, F. J., Baker, A. J. M., (2004). Research priorities for conservation of metallophyte biodiversity and their potential for restoration and site remediation. *Restor Ecol*. 12, 106-116.

Williamson, S. D. Balkwill, K., (2006). Factors determining levels of threat to serpentine endemics. *S. African J. Bot*. 72, 619–626.

Wuana, R. A., Okieimen, F. E., (2011). Heavy metals in contaminated soils: A review of sources, chemistry, risks and best available strategies for remediation. *ISRN Ecology*. 1-20.

Xiong, J., He, Z., Liu, D., Mahmood, Q., Yang, X., (2008). The role of bacteria in the heavy metals removal and growth of *Sedum alfredii* Hance in an aqueous medium. *Chemosphere*. 70, 489-904.

Xu, J., Tian, Y. S., Peng, R. H., Xiong, A. S., Zhu, B., Jin, X. F., Gao, J. J., Hou, X. L., Yao, Q. H., (2009). Yeast copper-dependent transcription factor ACE1 enhanced copper stress tolerance in Arabidopsis. *BMB Rep*. 42, 752-757.

Zhao, F. J., Lombi, E., Breedon, T., McGrath, S. P., (2000). Zinc hyperaccumulation and cellular distribution in *Arabidopsis halleri*. *Plant Cell Environ*. 23, 507-514.

Zhu, Y. L., Pilon-Smits, E. A. H., Tarun, A. S., Weber, S. U., Jouanin, L., Terry, N., (1999). Cd tolerance and accumulation in Indian mustard is enhanced by overexpressing g-glutamylcysteine synthetase. *Plant Physiol*. 121, 1169-1177.

In: Brassicaceae
Editor: Minglin Lang

ISBN: 978-1-62808-856-4

Chapter 7

THREE-DIMENSIONAL MOLECULAR STRUCTURE PREDICTION OF SELENOCYSTEINE METHYLTRANSFERASE (BOSMT) FROM *BRASSICA OLERACEA*

Raman Chandrasekar,[1,*] P. G. Brintha,[2] Minglin Lang,[1] M. Chandrasekaran[3] and K. Murugan[4]

[1]Department of Biochemistry and Molecular Biophysics, Biotechnology Core Facility, Kansas State University, Manhattan, US
[2]Department of Biology, Kansas State University, Manhattan, US
[3]Department of Environmental and Biological Chemistry, College of Agriculture and Life Sciences, Chungbuk National University, Chungbuk, South Korea
[4]Department of Zoology, Bharathiar University, Coimbatore, Tamilnadu, India

ABSTRACT

Recent attraction and follow-up research on the genomic and biochemical studies of *Brassica oleracea* reveal the importance of BoSMT genes that trigger the cascade of responses leading to Selenium (Se) accumulation in model plants. On the other hand bioinformatics plays an essential role in today's plant science. As the amount of data grows exponentially, there is a parallel growth in the demand for tools and methods in data management, visualization, integration, analysis, modeling, and prediction. Structure prediction is used both for assessing the quality of the newly determined structure and predicting the structure of proteins whose (BoSMT) sequences are newly determined. The necessary condition for successful homology modelling is to have sufficient similarity between the protein sequences. Here we construct the BoSMT enzyme protein structure based on the *Thermotoga maritima* (Tm-HMT) by using MODELER program. From Ramachandran plot analysis protein residues falling into the most favoured regions were determined (83.3%). The predicted molecular 3D model further verified with PROCHECK, VADAR online server to confirm the geometry and stereo-chemical

* Tel. 785-532-6125, Email: biochandrus@yahoo.com, chandbr@ksu.edu.

parameters of molecular architecture. Successfully modelled, verified and the most reliable structure of BoSMT was used for deposition in PMDB (Protein Model Database) database accession No. PM0078717. The multiple sequence alignment of BoSMT with related plants showed high sequence homology (87% BoHMT, 72% *Malus domestica*, 70% *Astragalus chrysochlorus*, 69% *A. racemicus* and *A. drummondii*, 68% *Camella sinesis*, 65% *Arabidopsis thaliana* (AtHMT2), 53% Sorghum, 38% *Zea mays*). Moreover active binding site and functional implication were discussed for BoSMT enzyme.

Keywords: Active site, selenium, selenocysteine methyltransferase, *Barrisca oleracea*, 3D homology, homocysteine methyltransferase, Ramachandran plot

Introduction

Broccoli (*Brassica oleracea* var. italica) is a common vegetable consumed worldwide (Soengas *et al*., 2011). Broccoli has been described as a 'super–vegetable' among consumers after the numerous epidemiological and laboratory studies on this *Brassica* species. *Brassica* species in comparison with other vegetables have high antioxidant capacity which makes them a very attractive crop from the consumer points of view. Broccoli has the ability to accumulate high level of Se-methylselenocysteine (SeMCys) and SeMet when grown on seleniferous soil.

A recent review provides information on the presence of selenium content in food, its associated health effects, technical approaches used for speciation (Dumont *et al*., 2006; Rayman *et al*., 2008; Pedrero *et al*., 2009; Susan *et al*., 2010). Selenium (Se) has been studied extensively due to its beneficiary effects in animals and humans. These seleno aminoacids has been on clinical trials and epidemiology studies have shown that cruciferous vegetables consumption, broccoli in particular, reduces the risk of several chronic diseases, such as cardiovascular diseases, inflammation, aging–related disorders and certain types of cancer (Ip *et al*., 2000; Whanger 2002; Traka *et al*., 2008; Bjorkman *et al*., 2011). The molecular mechanism of cancer prevention by selenium using the genomics approach was studied on the target organ breast, prostate, colon and lung. The result of the microarray analysis indicated that selenium, independent of its form and the target organ, alters several genes in a manner that can account for cancer prevention.

Broccoli is initially absorbing selenium from soil and converts it to selenium containing amino acids such as Se-methylselenocysteine and selenomethionine. The biosynthesis of most selenium compounds in nature follows the pathways leading to isologous sulfur compounds in plants (Table 1 and Figure 1). Se-methylselenocysteine is a naturally occurring L-form selenoamino acid found in plants of the *Brassica* families. As much as 80% of the total selenium found in Broccoli is present as Semethylslenocysteine (Whanger, 2002; White *et al*., 2007; Pilon-Smits and Quinn, 2010). The selenium content of Broccoli is very high (62.3^2 μg/g fresh weight) depending on soil content and other environmental factors, whereas in other plant food the content is generally lower (Susan *et al*., 2010). Increasing evidence showed that Se-accumulating plant species are known to express selenocysteine methyltransferase-SMT (Sors, Martin and Salt 2009). In particular broccoli has the capacity to convert selenocysteine to Se-methylselenocysteine, it is not yet clear whether the previously cloned broccoli SMT (Lyi *et al.*, 2005) is indeed a *bona fide* SMT or rather a homocysteine methyltransferase with some SMT activity (Lyi *et al.*, 2005; Sors *et al.*, 2005).

Table 1. Selenium-containing compounds in plants (adopted from Birringer *et al.*, 2002)

Compounds	Plant species
Selenocysteine	*Astragalus praleongus* *Astragalus pectinatus* *Brassica oleracea capitata* *Lecythis ollaria* *Morinda reticulate* *Neptunia amplexicaulis* *Stanleya pinnata* *Vigna radiata*
Se-methylselenocysteine	*Astragalus crotalariae* *Astragalus biculcatus* *Astragalus praleongus* *Allium sativum* *Allium cepa* *Allium tricoccum* *Brassica oleracea capitata* *Brassica oleracea botrytis* *Melilotus indica* *Oonopsis condensata* *Phaseolus lunatus*
ϒ-Glutamyl-Se-methylselenocysteine	*Astragalus bisulacatus* *Allium sativum* *Allium cepa* *Phaseolus lunatus*
Selemethionine	*Allium tricoccum* *Brassica juncea* *Brassica oleracea* *Melilotus india*
Se-methylselenocyteine Se-oxide	*Brassica oleracea capitata*
Selenobiotin	*Phycomyces blakesleeanus*
ϒ-Glutamylselenocystathionine	*Astragalus pectinatus*
ϒ-Glutamylselenomethionine	*Allium sativum*
3-Butenyl isoselenocyanate	*Stanleya pinnata*
Selenosinigrins	*Armoracia lapathifolia* *Stanleya pinnata*
Selenosugars	*Astragalus racemosus*

The contradictory findings proved that the BoSMT lacks obvious chloroplast or mitochondrial targeting sequences and appears to be a cytosolic enzyme (http://hc.ims.u-tokyo.ac.jp/iPSORT/).

Although the cytosolic location of BoSMT is yet to be confirmed in the plant, such location is consistent with other published data showing that methylation of selenocysteine or selenomethionine as well as metabolism involving the synthesis of S-methylmethionine most likely takes place in the cytosol (Bourgis *et al.*, 1999; Ranocha *et al.*, 2000).

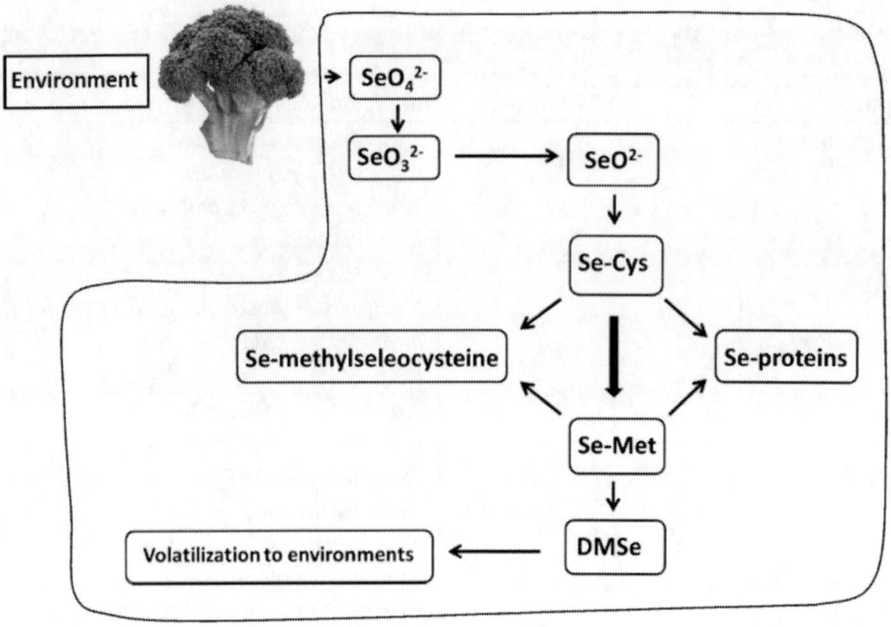

Figure 1. Schematic overview of Se metabolism in plants. (Adopted and modified from Germ *et al.*, 2007).

Furthermore, the BoSMT enzyme shares over 40% homology to YagD, a homocysteine methyltransferase enzyme from *E.coli* (Neuhierl *et al.*, 1999), suggesting that SMT and HMT enzyme (both at the nucleotide and amino acids levels) share a common ancestry and are functionally related. Interestingly, the Tm-HMT (3Bof) proteins is well conserved unlike all plant-derived HMT sequences as well as the putative SMT sequences isolated from broccoli (BoSMT; Lyi *et al.*, 2005) and *Camellia sinensis* (CsSMT; Zhu *et al.*, 2008).

Se-methyltransferase have also been proposed to exhibit a cancer-preventive potential when consumed regularly in the human diet (Keck & Finley 2004; Finley *et al.*, 2001, 2005; Verkerk *et al.* 2008; Hasanuzzaman and Fujita, 2011; El Mehdawi *et al.*, 2011). Se-methyltransferase enzyme (micronutrients) is essential for normal physiological and metabolic processes of human beings. So far there is no NMR or X-ray protein structure of BoSMT enzyme available in PDB database. Hence attempt was made to design 3D molecular structure of Se-methyltransferase and to clearly demonstrate the evolutionary relationship between the SMT and HMT proteins in plants.

METHODOLOGY

Target Protein Sequence and Template Selection

The protein sequence of BoSMT from *Brassica oleracea* was obtained from the protein sequence database of NCBI (Accession No. DQ67980.1). It was ascertained that the three dimensional structure of the protein was not available in Protein Data Bank (http://www.rcsb.org/pdb). The NCBI BLAST was used to identify the template for modeling the three dimensional structure of *Thermotoga maritima* (3Bof) from bacteria and *Escherichia coli* (YagD). The result of NCBI BLAST against the PDB database was used for selection of a suitable template for 3D modeling of the target protein.

Sequence Alignment

BoSMT amino acid sequence was used for alignment with template protein using PSI-BLAST (http://blast.ncbi.nlm.nih.gov/Blast). Default parameters were applied and the aligned sequences were inspected and adjusted manually to minimize the number of gaps to better understand the functional behavior of this protein. An *in-silico* study, mainly comparative homology modeling, of the target sequence BoSMT can be helpful to investigate sequential-structural-functional relationship. 3D structure of BoSMT was predicted based on available homologous template structure in Protein structure Data Bank (PDB) resources. Template selection was performed using PDB advance BLAST (http://www.rcsb.org). Retrieved template structure was used for comparative homology modeling of BoSMT.

Phylogenetic Tree Analysis

Several programs exist for making phylogenetic trees that display the relationship between sequences to calculate all possible tree topologies to find the one that fits best with the sequence data. We used Clustal X (Thompson *et al.*, 1997) to generate alignments of the sequences. Nucleotide/protein sequences were subsequently aligned manually in order to increase alignment. A bootstrapped, unrooted Neighbor-Joining tree was generated (pairwise) using the same program. After getting the root distance format using the Clustal W, the code was submitted into the phylo-draw software version 8.2 in NJ plot. Further polypeptide hydropathy/amphipathicity was evaluated using the Kyte–Doolittle algorithm (Kyte and Doolittle, 1982) as implemented in the GCG computer program (Genetics Computer Group, Wisconsin Package version 8.1, Wisconsin). A window of nine residues was used in the analysis. The GCG program was run on a Macintosh computer with eXodus 5.2 software (White Pine Software, Inc.). The hydropathy/amphipathicity plot obtained from the GCG program was saved as text form and redrawn using Kaleida- Graph software.

Homology Modeling and Structure Refinement

The three dimensional structure of BoSMT has been predicted using DS MODELLER (http://salilab.org/modeller) and SWISS MODEL (http://swissmodel.expasy.org). A rough 3D model was constructed based on sequence alignment between 3Bof of bacteria with parameters of energy minimization value. From the homology modeling searching, two templates were selected - High-resolution X-ray crystallography structure of the 3Bof of *T. maritima* (Evans *et al.*, 2004; Koutmos *et al.*, 2008) and YagD (Bernhard Neuhierl *et al.*, 1999).

Out of the two models, 3Bof was chosen to the best model according to the scoring of Procheck, total non-local energy of the protein (*E/kT* units) and overall model quality Z-score. Loop refinement and structural simulation were done using LOOPER and CHARMm force field, respectively. Finally, predicated 3D model was subjected to a series of tests for testing its internal consistency and reliability.

The quality of model was checked, verified 3D (Eisenberg *et al.*,1997), Profile 3D (Suyama *et al.*, 1997) and Errat (Gundampti *et al.*, 2012) and the stereo-chemical properties based on backbone conformation were evaluated by inspection of Psi/Phi/Chi/Omega angle using Ramachandran plot of MolProbity (http://molprobity.biochem.duke.edu/). Further quantitative analysis was done using accessible surface area prediction using Volume Area Dihedral Angle Reporter (Table 2, 3, 4) by using VADAR (http://vadar.wishartlab.com/) online server (Willard L *et al.*, 2003).

Table 2. VADAR statistics report. Secondary elements details with hydrogen bond statistics*

VADAR STATS using atomic radii from sharke		
Statistic	Observed	Expected
I. Secondary structure		
# Helix	144 (41%)	-
# Beta	71 (20%)	-
# Coil	131 (37%)	-
# Turn	48 (13%)	-
II. Hydrogen bands		
Mean hbond distance	2.3 sd=0.4	2.2 sd=0.4
Mean hbond energy	-1.6 sd=1.1	-2.0 sd=0.8
# res with hbonds	269 (77%)	259 (75%)

* The expected values represent those numbers which would be expected for highly X-ray and NMR structures.

Table 3. Dihedral angles observation based on phi, psi, and chi and omega calculations #

Statistic	Observed	Expected
Mean Helix Phi	-64.0 sd=6.1	-65.3 sd=11.9
Mean Helix Psi	-40.9 sd=16.1	-39.4 sd=25.5
# res with Gauche+ Chi	114 (41%)	152 (55%)
# res with Gauch- Chi	46 (16%)	55 (20%)
# res with Trans Chi	118 (42%)	69 (25%)
Mean Chi Gauche+	-68.0 sd=11.1	-66.7 sd=15.0
Mean Chi Gauch-	63.8 sd=11.8	64.1 sd=15.7
Mean Chi Trans	170.7 sd=8.7	168.6 sd=16.8
Std. dev of Chi pooled	10.23	15.70
Mean omega (omega >90)	-178.9 sd=5.0	180.0 sd=5.8
# res with (omega <90)	0	0

\# Expected values obtained from Morris Al, MacArthur MW, Hutchinson EG and Thornton JM. Proteins. 1992, 12 (4): 345-364.

Structure Submission

Successfully modeled, verified and the most reliable structure of BoSMT was used for deposition in PMDB (Protein Model Database) database accession No. PM0078717 (http://mi.caspur.it/PMDB/).

Active Binding Site Prediction

After complete modeling, simulation and refinement of the structure of BoSMT (PMDB ID: PM0078717), prediction of the possible metal binding site was performed using Q-site Finder (http://bmbpcu36.leeds.ac.uk/qsitefinder/). PyMOL software used to illustrate the binding site of the target protein. These binding sites were further compared to the active sites of the template.

RESULTS AND DISCUSSION

An increasing amount of data from different "omics" projects is present worldwide. Therefore the need for an efficient and powerful biological data management system, providing trends and relationship among the stored data, has become quite obvious. Bioinformatics was developed to handle the large amount of genetic and biomedical data (Mount, 2001), which reflect the sequence, structure and functions hierarchy enabling the connection of different information in their natural and medical context. The success and applicability of homology modeling is steadily increasing due to growing number and availability of experimentally determined protein structures (Macro Wiltgen and Tilz, 2009). The selected protein BoSMT enzyme (Accession No. DQ67980.1) has 346 amino acid in length with molecular weight of ~ 37.8 kDa. It is >40 % of Leu, Ala, Ser, Glu, rich amino acid with theoretical pI 5.41. Earlier studies by Sors *et al.*, (2009) showed the MW of *Astragalus* species 37.6 kDa. Out of 346 amino acid residues Serine was found to be highest number with 33 residues (9.5%), followed by Alanine, Leucine and Glutamic acid 30 residues (8.7%) and second highest Glycine and Isoleucine with 27 residues (7.8%). The molecular formula of BoSMT was found to be $C_{1681}H_{2656}N_{448}O_{532}S_{11}$ with a total number of 5319 atoms. The instability index was computed to be 38.77 which classified the protein as stable (Guruprasad *et al.*, 1990).

The BoSMT enzyme belongs to a class of methyltransferase involved in metabolism of S-methylmethionine. It shared significant primary sequence homology with homocysteine S-methyltransferase (87% BoHMT) from *B. oleracea* as well as (65%) Arabidopsis (Ranocha *et al.*, 2000). However both BoSMT and BoHMT catalyze methyl transfer using S-methylmethione as the methyl donor, they exhibit remarkable Se-containing (SMT) and S-containing (HMT) substrate preference as a methyl acceptor *in vitro* (Neuhierl and Bock, 1996; de Souza *et al.*, 2000; Ranocha *et al.*, 2000). Phylogenetic analysis of BoSMT and related SMT/HMT homologs are shown in Figure 2. The un-rooted phylogenic tree result suggest an independent origin of BoSMT enzymes, with both could have been derived from common ancestor to evolutionary group of microorganism (bacteria), which is in agreement with similar conclusions reached in recent independent study (Iyar *et al.*, 2011). The BoSMT clade comprises several clearly defined subgroups, their plant BoHMT counterparts. In a similar way, BoSMT and BoHMT, whose corresponding genes are organized in tandem, are localized in the same evolutionary branch, together with representative proteins from other different species. BoSMT-related proteins indicate a relatively early diversification of BoSMT families during plant evolution. Within the family of BoSMT and BoHMT enzymes, multiple sequence analysis with other related plants revealed that the N- and C-terminal halves exhibit significant sequence homology (Figure 3). The C-terminal half is more highly conserved than the N-terminal half and homologs with 87% BoHMT, 72% *Malus domestica*, 70% *Astragalus chrysochlorus*, 69% *A. racemicus* and *A. drummondii*, 68% *Camella sinesis*, 65% *Arabidopsis thaliana* (AtHMT2), 53% *Sorghum sp.*, 38% *Zea mays*.

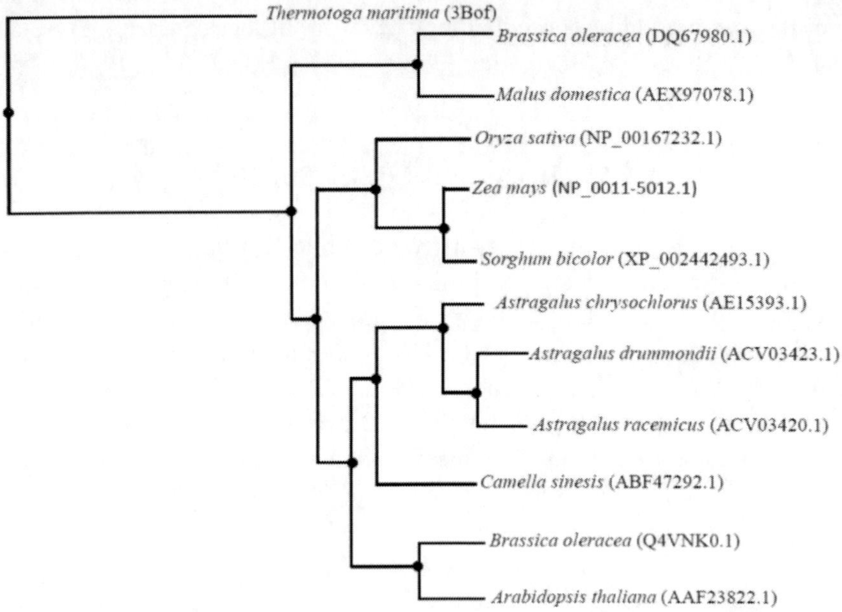

Figure 2. Phylogenetic analysis of the aligned protein sequences. The analysis was conducted using the branch and bound parsimony algorithm and single un-rooted phylogram tree was identified. The length of the branch lines indicates the extent of divergence according to the scale 2.0 bar. (Abbreviation and accession number more detail in Figure 3).

On the basis of sequence similarity analysis, *Brassica oleracea* showed 54% sequence similarity and 40% with template structure (PDB ID: 3Bof and YagD) respectively. Since the template showed a good level of sequence identity it was used to obtain high quality alignment based structure prediction using homology modeling. The output of the PSI-BLAST program is a list of alignments of the query sequence with different potential homologous sequence. The alignment with low expectation-value (*E-value*) is very significant, that means there is a high probability that the sequences are homologous (Figure 4). A PDB ID: 3Bof X-ray crystal structure of *Thermotoga maritima* (3Bof) from bacteria was specifically selected on the basis of BLAST result and was utilized as a template for structure modeling of BoSMT. The sequence of the target and the template are brought into an optimal alignment (Figure 4a).

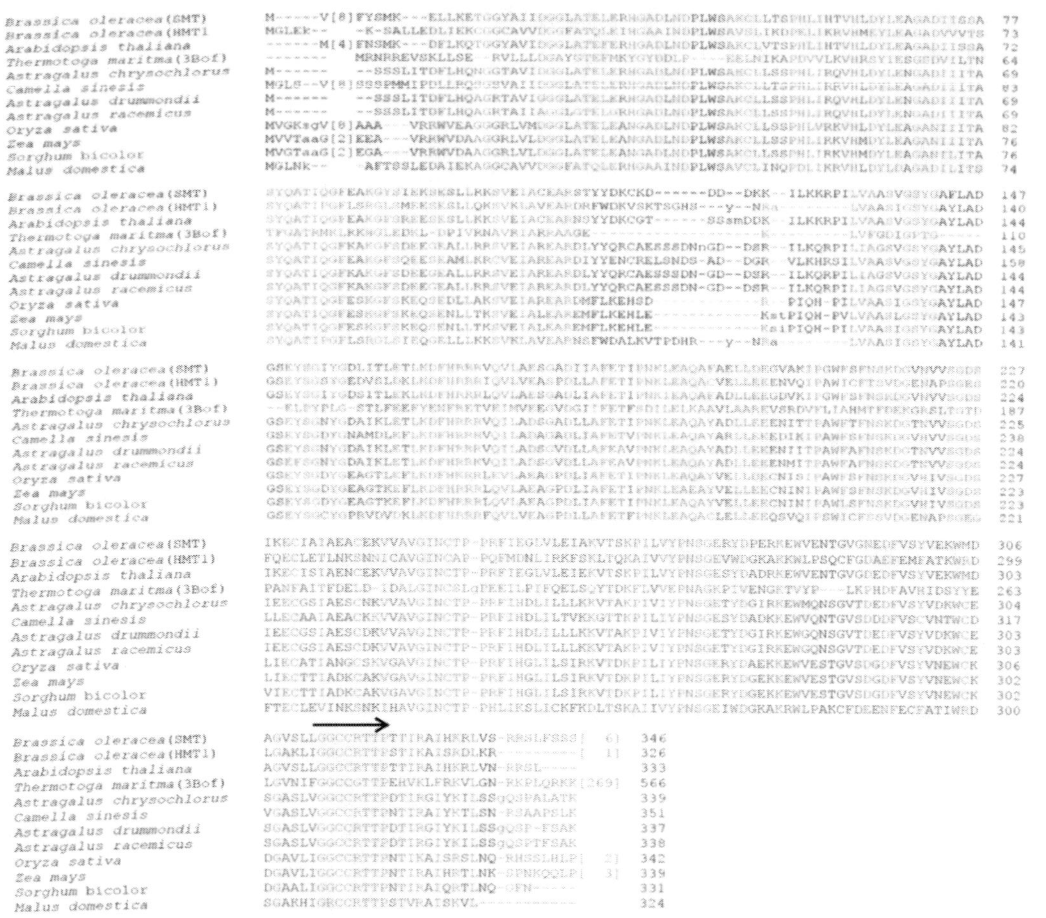

Figure 3. Multiple sequence alignment of the deduced amino acids for *Brassica oleracea* (BoSMT) with those of related other plants. Alignment of amino acid sequences of BoSMT (DQ667980.1) with related proteins from different organisms, including *Oryza sativa* (NP_00167232.1), *Malus domestica* (AEX97078.1), *Zea mays* (NP_0011-5012.1), *Sorghum* bicolor (XP_002442493.1), *Astragalus chrysochlorus* (AE15393.1), *Astragalus drummondii* (ACV03423.1), *Astragalus racemicus* (ACV03420.1), *Camella sinesis* (ABF47292.1), *Arabidopsis thaliana* (AAF23822.1) and *Brassica oleracea*-HMT (Q4VNK0.1). Highly Conserved residues are marked as light gray and dark black color and black arrow shows the possible zinc-binding motif.

An optima alignment refers to considerable number of positive or negative identical (that means the residue can be easily substituted by similar residues, negative means the residues can be hardly substituted) corresponding residues and only little gaps. Then, the structure of the target protein is constructed by exploiting the information from the template structure (Figure 4b). The modeling steps are: backbone superposition of the atom; loop modeling and orientation of the side chains (Figure 5a-d). There are several ways to model the backbone (Sali and Blundell, 1993; Sanchez and Sali, 1997). The target backbone (from N-terminus to C-terminus is built by averaging the backbone atom position of the template structures. Structural model was built based on the atomic coordinate in the PDB data files, the structure are visualized in the appropriate representation (Figure 5d). The loop refined model, which was selected with minimum CHARMm energy (40639.47232 kcl/mol) based on conjugant gradient minimization, was considered to evaluate qualitatively and quantitatively.

a

```
BoSMT     MVTGNTKAETFYSMKELLKETGGYAIIDGGLATELERHGADLNDPLWSAKCLLTSPHLIH 60
3Bof      -------MPNRPEVSKLLSER--VLLLDGAYGTEFMKYGYDD----LPEELNIKAPDVVL 47
                 ..  .:.:**.*        ::**. .**: ::* *        . :  :.:*.::

BoSMT     TVHLDYLEAGADIISSASYQATIQGFEAKGYSIEKSESLLRKSVEIACEAR---------- 111
3Bof      KVHRSYIESGSDVILTNTFGATPMKLRKHG-LEDKLDPIVRNAVRIARRAAGEKLVFGDI 106
          .** .*:*:*:*:* : :: **     :. :*    :* :.::*::*.** .*

BoSMT     ---------------STYYDKCKD---------------DDDKKILKKRPILVAASVGSY 141
3Bof      GPTGELPYPLGSTLFEEFYENFPETVEIMVEEGVDGIIFETFSDILELKAAVLAAREVSR 166
                     . :*:: ::               :  ..**: :. ::**    *

BoSMT     GAFLADGSEYS-----------------------------GIYGDLITLETLKDEHR 169
3Bof      DVFLIAHMTFDEKGRSLTGTDRANFAITFDELDIDALGINCSLGPEEILPIFQELSQYTD 226
          ..**    :.                               *     *    :: *.::

BoSMT     RRVQVLAESGADIIAFETIPNKLEAQAFAELLD-----EGVAKIPGWFSFN----------- 215
3Bof      KFLVVEPNAGKPIVENGKTVYPLKPHDFAVHIDSYYELGVNIFGGCCGTTPEHVKLFRKV 286
          : : * .::* *:   .     *:.: **  :*       ** : *  . .

BoSMT     ---------SKDGVNVVSGDS-------------------------------------IKEC 231
3Bof      LGNRKPLQRKKKRIFAVSSPSKLVTFDHFVVIGERINPAGRKKLWAEMQKGNEEIVIKEA 346
                   .*. : .**. *                                  ***.

BoSMT     IAIAEACEKVVAVGINCT---PPRFIEGLVLEIAKVTSKPILVYPNSGERYDPERKEWVE 288
3Bof      KTQVEKGAEVLDVNFGIESQIDVRYVEKIVQTLPYVSNVPLSLDIQNVDLTERALRAYPG 406
           : .*  :*: *.:.        *::* :*  :. *:. *: :  :. :  :   : :

BoSMT     N------TGVGNEDFVSYVEKWMDAG----VSLLGGCCRTTP---------TTIRAIHKRLV 331
3Bof      RSLFNSAKVDEEELEMKINLLKKYGGTLIVLLMGKDVPKSFEERKEYFEKALKILERHDF 466
          .   : *.:*::   ::   . *     * *:*   .:           .::: :.:: .

BoSMT     SRRSLFSSSSSSSHH------------------------------------------- 346
3Bof      SDRVIFDPGVLPLGAEGKPVEVLKTIEFISSKGFNTTVGLSNLSFGLPDRSYYNTAFLVL 526
          * * :*...   .

BoSMT     ---------------------------------------
3Bof      GISKGLSSAIMNPLDETLMKTLNATLVILEKKELPRAEVK 566
```

b

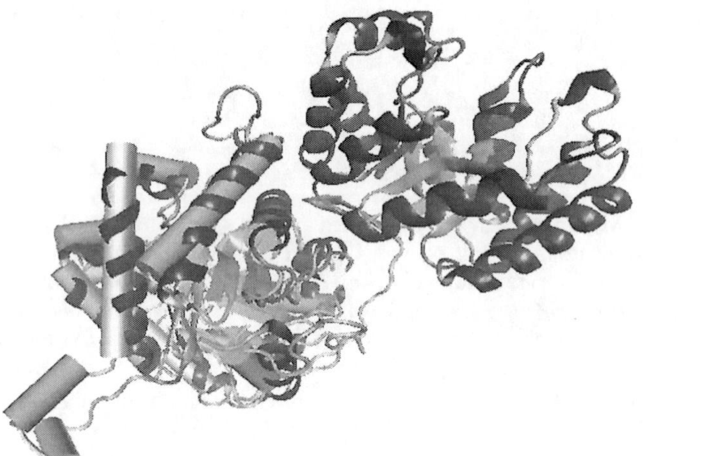

Figure 4. a) Se-methyyltransferase sequence of *Brassica oleracea* (BoSMT) alignment with cobalamin-independent methionine synthase enzyme of *Thermotoga maritima* (Tm-Meth), conserved residues are marked as *. b) 3D view shows the overlapping of BoSMT (cylindrical model) with selected template (ribbon model).

3D predicated model was analyzed using energy minimization, refinement and simulation program of PROCHECK. PDB Sum server was employed for evaluation by comparing the geometry and stereochemical parameters (r_0, Θ and ϕ) quality of predicted model (Figure 6a, b). A large number of literature related homology modeling were also found to use the PROCHECK for screening the best model (Kherraz *et al.*, 2011; Sharma and Bhatacharya, 2012 and Neha Arora *et al.*, 2012). The resulting structures (1.8 – 2.0A°) can be visualized by using PyMOL program in several representations such as line drawing (Figure 5 and Figure 7), balls-and-sticks, cylindrical, ribbon model (Figure 7a), transparent grid surface of ribbon model (Figure 7b), space fill model (Figure 7c) (showing secondary structure elements) and the molecular surface (Figure 7d) can be calculated. It has been from the BoSMT structure that the helix and the beta sheet regions of the template and model structure were superimposed in a better way compared to the loop regions (Figure 5c). It has been known from the literature that loop regions are main region, where accuracy of a model protein structure deviates from the template (Al Lazikani *et al.,* 2001; Fisher *et al.,* 2002). Further Ramachandran plot can be used to evaluate how well the values of the dihedral angles agree with the values of allowed conformation for protein backbones (Figure 6c). Ramachandran plot analysis showed 83.3% of amino acid residues within the most favoured and 15% residues in additional and generously allowed regions, whereas 4 residues were found in disallowed region (Figure 8a, b). The comparable Ramachandran plot characteristic and G-factor score confirmed the good quality of the present predicated model. Based on main-chain and side-chain parameters study, we found that the confirmation of the predicated model was very much favorable, stable and 99.9% accurate.

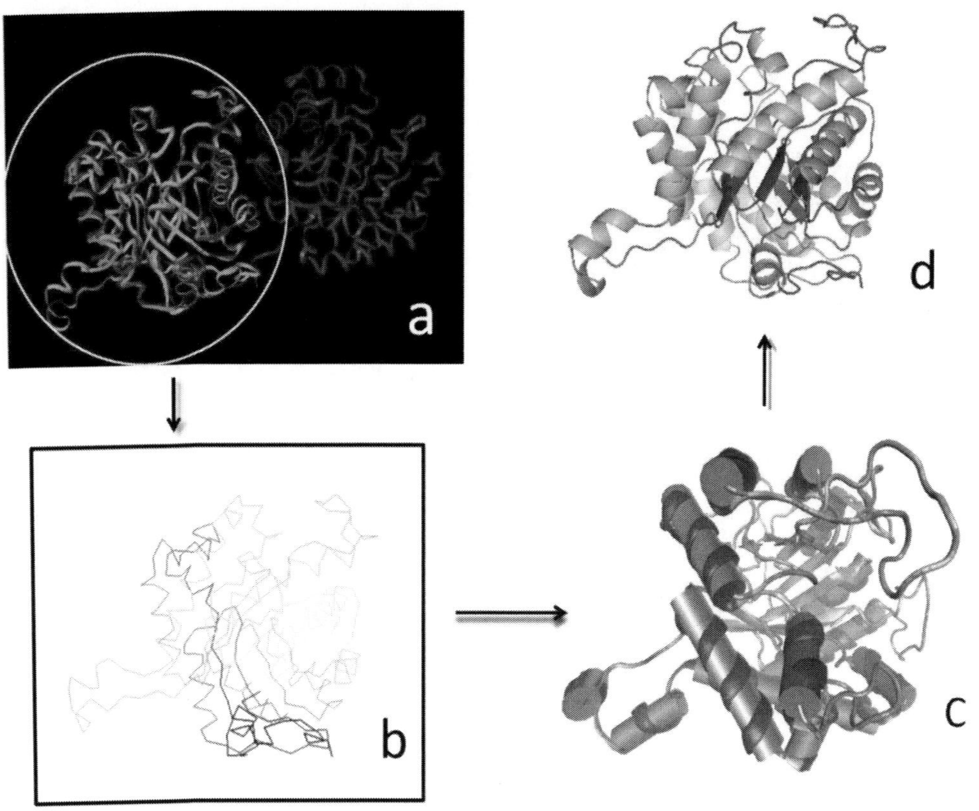

Figure 5. Deduced 3D structure of *Brassica oleracea* (BoSMT) predicated using the cobalamin-dependent methionine synthase from *Thermotoga maritima* (PDB-3Bof) as a template. a) Structure superposition of BoSMT (orange) with template (green); b) Line drawing for BoSMT; c) Distinguish structure superposition of BoSMT (Cylinder model) with 3Bof (Ribbon model); d) Predicated 3D homology modeling of BoSMT.

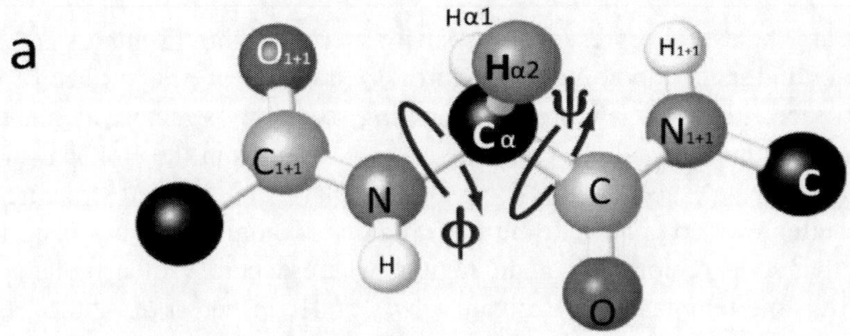

b **Ramachandran plot statistic**

Residues in most favoured region (A,B,L)	255	83.3%
Residues in additional allowed regions (a,b,l,p)	41	13.4%
Residues in generously allowed region (~a,~b,~l,~P)	6	2.0%
Residues in disallowed regions	4	1.3%
Number of non-glycine and non-proline residues	306	**100%**
Number of end-residues (excl.Gly and Pro)	2	
Number of glycine residues (shown as triangels)	27	
Number of proline residues	11	
Total number of residues	**346**	

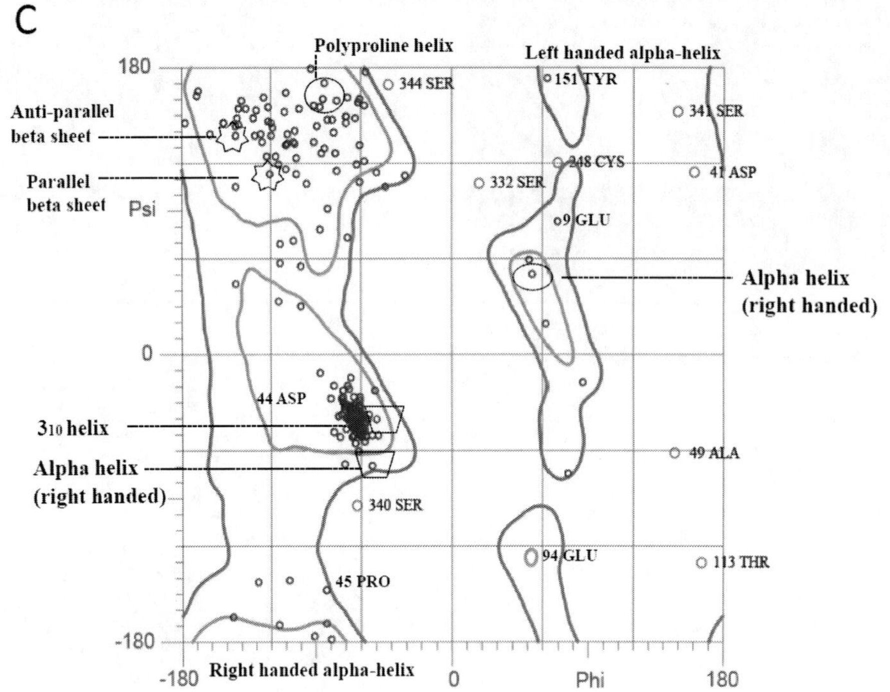

Figure 6. Ramachandran plot analysis of model using MolProbity software. a) Schematic representation of dihedral angles of ϕ and Ψ. b) Statistic value of Ramanchandran plot, c) Ramachandran plot shows how good the value of the dihedral angles agree with the value of allowed confirmation. The glycine residues (255 residues are in favoured region, 41 is allowed region and 4 in disallowed) 46+1 residues have allowed confirmation of BoSMT structure.

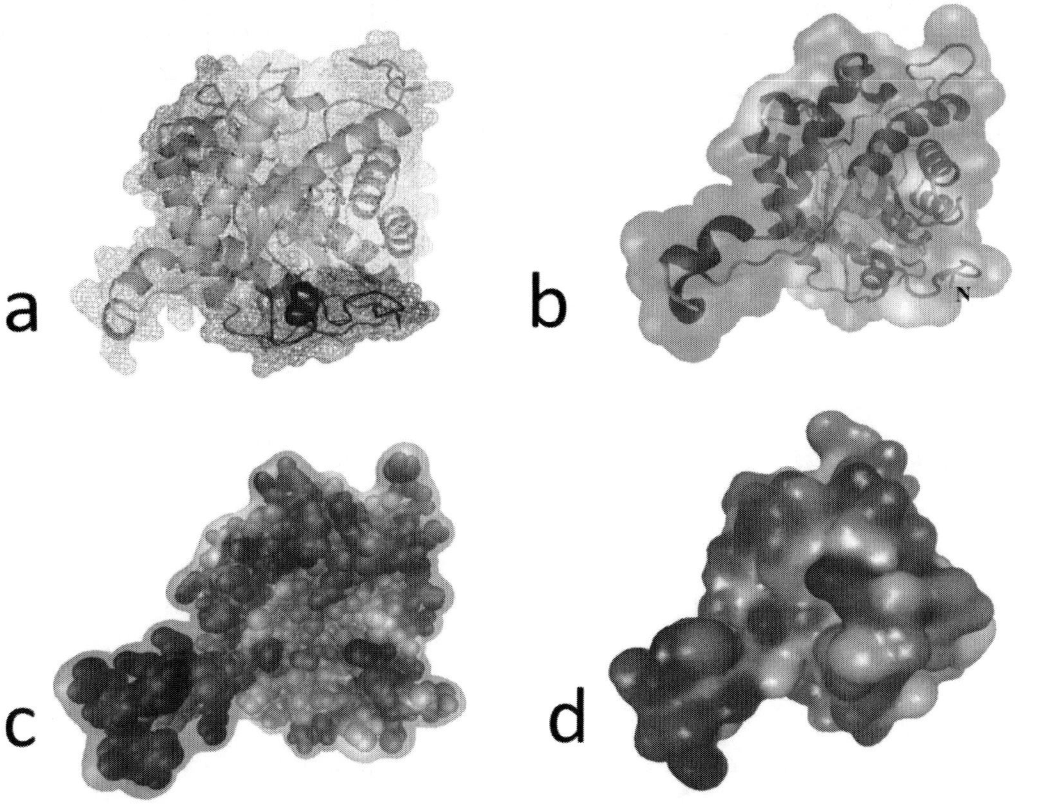

Figure 7. Different 3D molecular structure of BoSMT. a) Ribbon model with mesh surface; b) Transparent grid surface of Ribbon model; c) space fill colored atom view of hydrogenated; d) Stereo view of molecular surface model based on type of residue, color code: blue-charged amino acid, green-polar, white- hydrophobic, red-negative charge.

Hydropathy plot is a quantitative analysis of the degree of hydrophobicity or hydrophilicity of amino acids in a protein. Those inter helical contacts are mainly from hydrophobic residues such as leucine, valine, isoleucine, phenylalanine and alanine. These hydrophobic interactions are the major force contributing to stabilization of the helix bundle structure.

The hydrophobic residues have been highlighted quite interestingly in the multiple sequence alignments of various BoSMT. These residues are well conserved across all the sequences from the different plant species. Hence it is envisaged that a common mechanism may underlie in the folding and function of BoSMT. Furthermore, the results of an experiment by Xiong *et al.* (1995), showed that the hydropathic character of sequence residue has a larger effect on the sequence's choice for alpha-helix or beta-sheet, as compared to the intrinsic propensities of the amino acids for a particular secondary structure (Figure 8c).

However, further qualitative and quantitative analysis of predicted BoSMT model by using VERFIY 3D details lie between 0.01-0.74 representing the best verified and reliable model of BoSMT (Figure 9a, b, c).

Overall quality factor was calculated with ERRAT server (Table 2, 3, 4) and the model structure was found to have 95% quality factor. VADAR that included accessible surface area, excluded volume, backbone and side chain dihedral angles, secondary structure, hydrogen bonding partners, hydrogen bond energies, steric quality, solvation free energy as well as local and overall fold quality yielded good result (Figure 9d). Using atomic radii from Sharke method, we observed 41% residue were involved in the formation of helices, 20% in beta sheets, 37% in coils and 13% residues formed turns. Observed mean hydrogen bond (h-bond) distance and energy value were closely similar with expected values in hydrogen bond statistics.

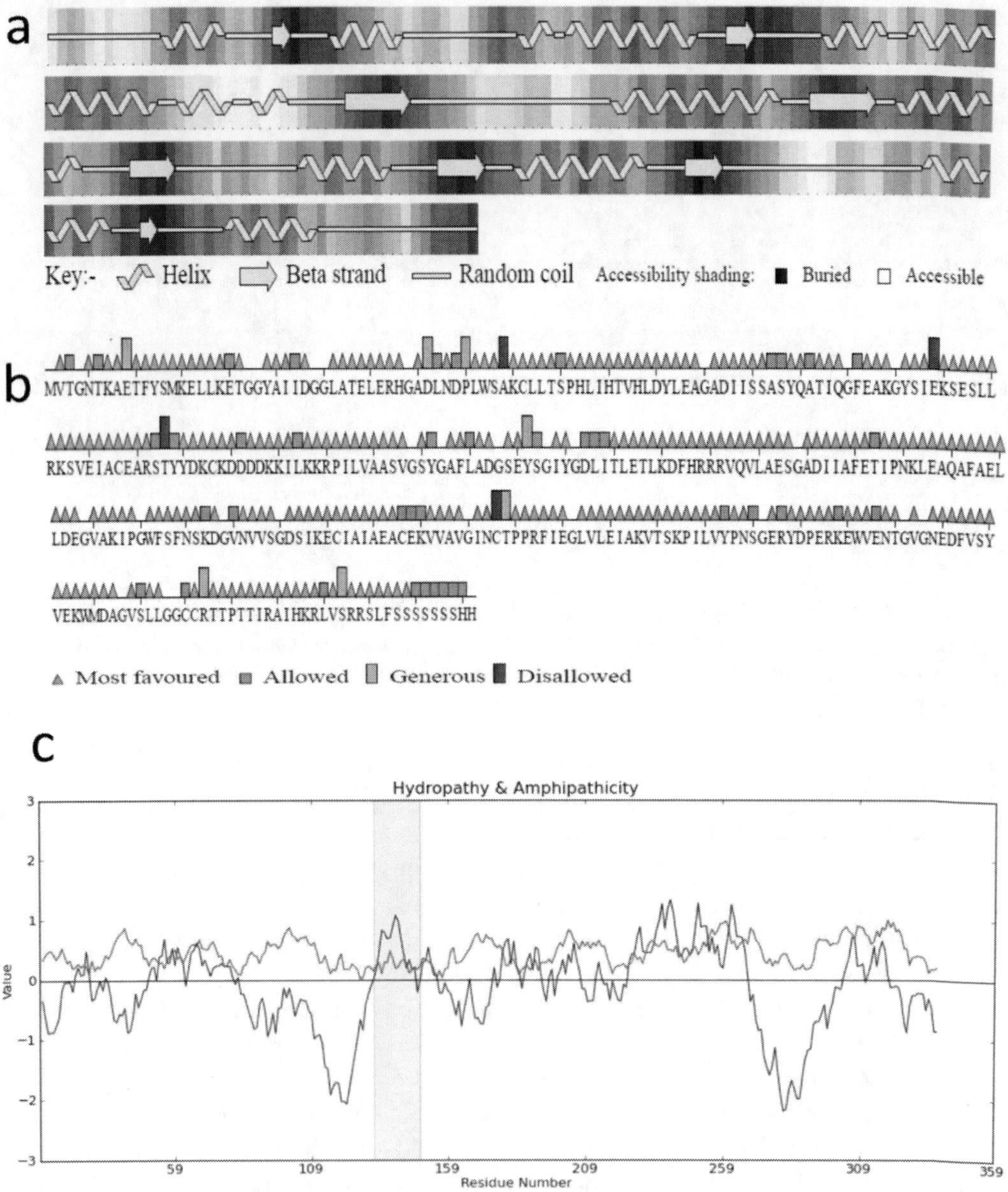

Figure 8. BoSMT sequence secondary predication and accessibility (a), 1D representation of residues found in most favoured allowed and generous allowed region (b), (c) Graph showed BoSMT hydropahty and amphipathicity. Red color denotes amphipathicity and Blue color denote hydropathicity.

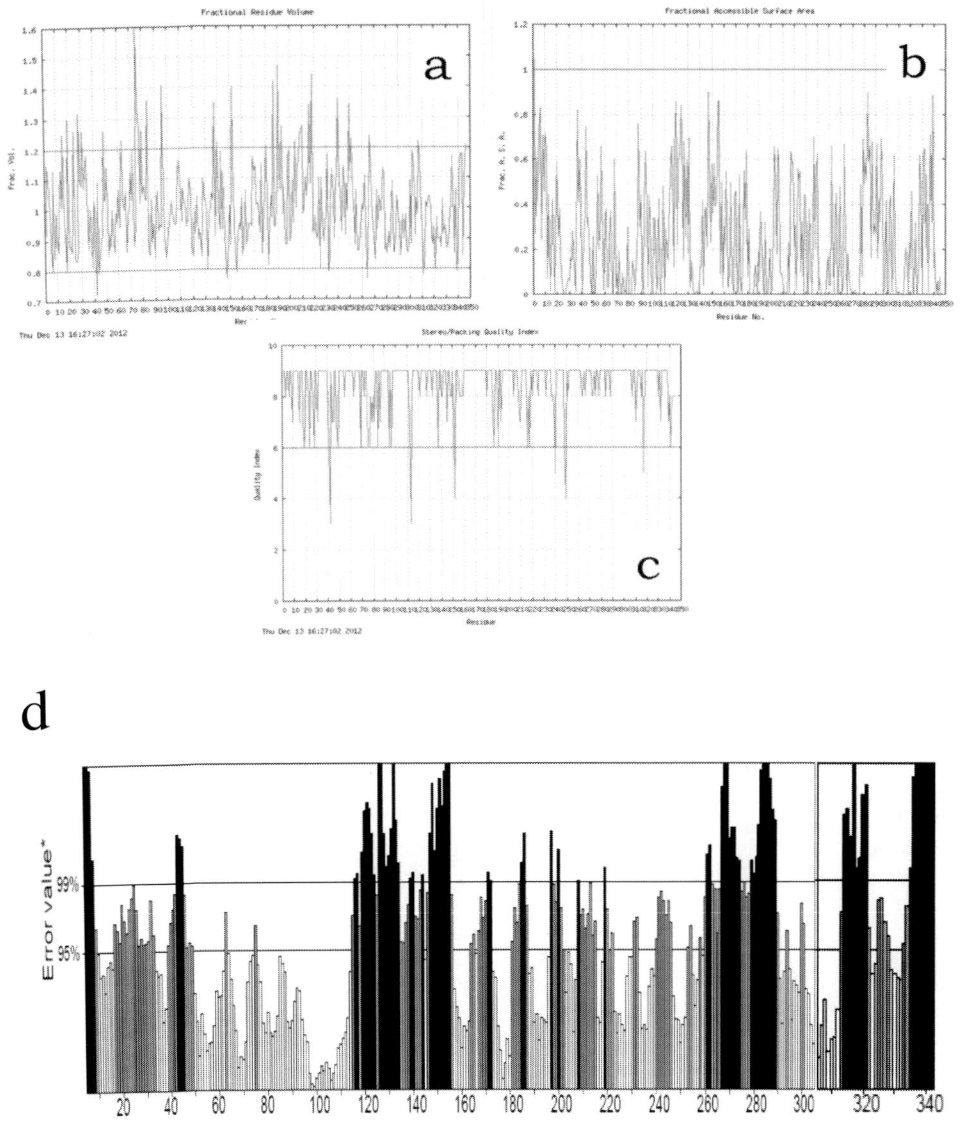

Figure 9. a) Factor residue volume, b) Factor A surface area and c) stereo/packing quality index of predicted model, d) Graph showed the quality factor obtained from ERRAT server.

The obtained expected residues with h-bond were 75% and we observed 77% for the predicted model. Dihedral angle statistics also represented approximately similar score with that of the expected values (Table 2, 3, 4). It was found that the overall quality and quantity on the basis of secondary elements of the predicated BoSMT model was good and reliable. The generated BoSMT model was successfully deposited in PMDB (http://www.caspur.it/PMDB) bearing Model ID: PM0078717. The resulting structure of BoSMT shows a homo-dimer with four stranded beta sheets surrounded by seven alpha-helices and four-stranded beta-sheets surrounded by three alpha-helices (Figure 10a, b, c).

Table 4. Qualitative report for accessible surface areas (ASA), accessible surface area for extended chain and their volume

Statistic	Observed	Expected
I. Accessible surface area (ASA)[a]		
Total ASA	15381.3 - Angs**2	13773.3 - Angs**2
ASA of backbone	1475.8 - Angs**2	-
ASA of sidechains	13905.4 - Angs**2	-
ASA of Carbon (C)	9037.5 - Angs**2	-
ASA of Nitrogen (N)	709.3 - Angs**2	-
ASA of Nitrogen (N+)	1180.6 - Angs**2	-
ASA of Oxygen (O)	3071.1 - Angs**2	-
ASA of oxygen (O-)	1261.1 - Angs**2	-
ASA of Sulfur (S)	121.8 - Angs**2	-
Exposed nonpolar ASA	8862.3 - Angs**2	9382.6 - Angs**2
Exposed polar ASA	2486.1 - Angs**2	3076.3 - Angs**2
Exposed charged ASA	4032.9 - Angs**2	2922.4 - Angs**2
Side exposed nonpolar ASA	8856.2 - Angs**2	-
Side exposed polar ASA	1055.3 - Angs**2	-
Side exposed charged ASA	3993.9 - Angs**2	-
Fraction nonpolar ASA	0.58	0.61 - sd=0.03
Fraction polar ASA	0.61	0.20 - sd=0.05
Fraction charged ASA	0.26	0.19 - sd=0.05
Mean residue ASA	44.5 - sd=47.2	-
Mean fraction ASA	0.3 - sd=0.3	-
% side ASA hydrophobic	23.57	-
II. Accessible surface area for		
extended coil [a]	36527.4 - Angs**2	-
Extended nonpolar ASA	16101.7 - Angs**2	-
Extended polar ASA	7401.0 - Angs**2	-
Extended charged ASA	36300.3 - Angs**2	-
Extended side non-polar ASA	3544.1 - Angs**2	-
Extended side polar ASA	7325.6 - Angs**2	-
Extended side charged ASA		
III. volume [b]	45020.6 - Angs**2	45813.6 - Angs**3
Total volume (packing)	130.1 - sd=38.4	125.0 - sd=40.0
Mean residue volume	1.0 - sd=0.1	1.0 - sd=0.1
Mean fraction volume	37864.21	-
Molecular weight (MW)		
IV. Resolution [c]	294 (84%)	311 (90%)
# res in phi-psi core	41 (11%)	24 (7%)
# res in phi-psi allowed	7 (2%)	3 (1%)
# res in phi-psi generous	2 (0%)	(0%)
# res in phi-psi outside	324 (93%)	332 (96%)
# res in omega core	17 (4%)	10 (3%)
# res in omega allowed	2 (0%)	(0%)
# res in omega generous	2 (0%)	3 (1%)
# res in omega outside	41	24
# packing defects	116	126
# res 95% buried	4	0
# buried charges	-322.10	-326.52
Free energy of folding		

a,b,c- expected value obtained from the J.Mol.Biol. 1987, 196 (3):641-656; Annu. Rev. Biophys. Bioeng. 1977, 6:151-176; Proc.Natl.Acad.Sci. USA, 1990, 87(8): 3240-3243.

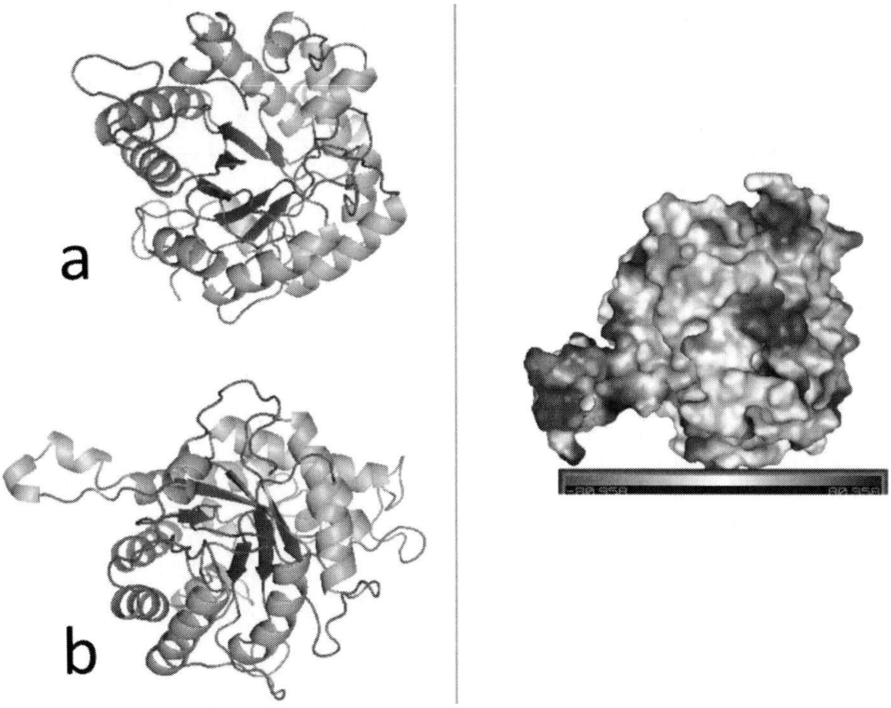

Figure 10. Structure was highlighted based on the secondary structure. a) top-view of the overall predicted BoSMT protein model (Helix-cyan, beta sheet-red and coil- pink); b) bottom-view ; c) the surface electrostatic potential of BoSMT protein (positive potential in blue and negative in red).

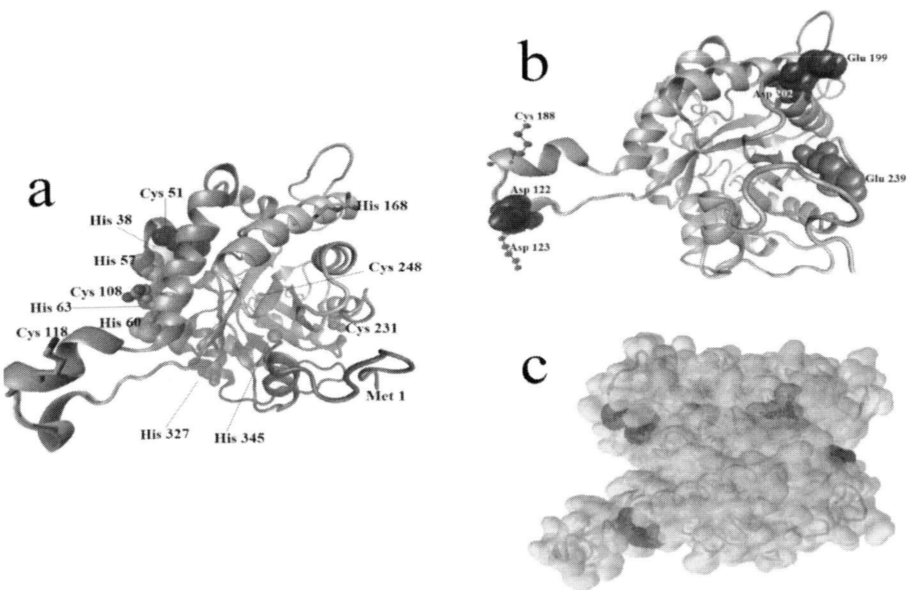

Figure 11. a) Detection of metal binding region, b) Ligand binding site and c) representing the same with surface model.

ACTIVE SITE FOR METAL BINDING REGION

Active site identification of 3D predicated model BoSMT with PMDB ID: PM0078717 from *Brassica oleracea* was done using Q-site finder. BoSMT contains a consensus sequence of GGCC for possible zinc-binding motif near the C-terminal and conserved Cys residue upstream of the zinc-binding motif as other related methyltransferase (Figure 11a).

Biological zinc site are usually four-coordinate with distorted tetrahedral geometry, although five and six-coordinate site have also been observed (Koutoms *et al.,* 2008). Common protein ligands for zinc centers are His, Glu^{199}, $Asp^{122-123}$, and Cys^{188}. This domain shares significant Se-sequence and structural homology with *Thermotoga maritime* (Pechal and Ludwing, 2005; Koutoms *et al*., 2008) and Arabidopsis (Ferrer *et al.,* 2004).

In SMT enzyme, the zinc site is situated near the top of a $(\beta\alpha)_8$ barrel and is assembled by residues perched at the C-terminal of inner barrel strands. Based on predicated binding site and the work conducted by others, we suggested that BoSMT may have zinc cofactor for binding and /or activating the selenol group of selenocystenine. Protein alignments of BoSMT and Tm-MetH confirmed the catalytic site of His^{38}, Cys^{51}, His^{57}, His^{60}, His^{63}, Cys^{108}, Cys^{118}, Cys^{168}, His^{168}, Cys^{231}, Cys^{248}, His^{327} and His^{345} residues. These were found to be prominent active binding sites for metal and protein-cofactors interactions (Figure 11b, c).

CONCLUSION

In the present work, a homology based 3D molecular structure of BoSMT from *Brassica oleracea* enzyme is constructed, using the MODELLER program. Precise evaluation and modeling of proteins is a major goal and key aspect of computational biology.

The resulting structural model is a three-dimensional representation of the protein that reflects empirical data in a consistent way by providing information about the spatial arrangement of groups of atoms. Being able to "see" the 3D structure of the BoSMT protein and analyzing its shape is of crucial importance for understanding protein properties and interactions.

It offers an alternative way to obtain structural information well before the structure of the new protein is determined by X-ray crystallography or NMR. These models offer the possibility to understand substrate recognition, specificity, by the analysis and visualization of the active sites. Further opened up new avenue for understanding the complete metabolism of Selenium. Interaction between selenium and other micronutrients, such as vitamin E, should be taken into consideration, particularly in relation to health outcomes that are associated with antioxidant nutrients for therapeutic approaches.

ACKNOWLEDGEMENTS

We would like to thank Dr. Gerald Reeck for constant encouragement and support. The authors are also thankful to Department of Biochemistry and Biotechnology Core Facility, Kansas State University for providing complete training and bioinformatics software. The authors had no personal or financial conflicts of interest.

REFERENCES

Al Lazikani, B., Jung, J., Xiang, Z., Honig, B., (2001). Protein structure prediction. *Curr. Opin. Chem. Biol.* 5, 51-56.

Bernhard Neuhierl, Martin Thanbichler, Friedrich Lottspeich, and August Bock, (1999). A Family of *S*-Methylmethionine-dependent Thiol/Selenol Methyltransferases : Role in selenium and evolutionary relation. *J. Biologial Chemistry*, 274 (9), 5407–5414.

Birringer, M., Pilawa, S., and Flohe, I., (2002) Trends in selenium biochemistry. *Natural Product Reports,* 19, 693-718.

Bourgis, F., Roje, S., Nuccio, M.L., Fisher, D.B., Tarczynski, M.C., Li, C. et al., (1999). S-methylmethionine plays major role in phloem sulfur transport and is synthezied by a novel type of methyltransferase. *Plant Cell*, 11, 1485-1497.

Bjorkman, M., Klingen, I., Birch, A.N.E., Bone, A.M., Bruce, et al., (2011). Phytochemicals of Brassicaceae in plants protection and human health-influences of climate environment and agronomic practices. *Phytochemistry*, 72, 538-556.

Dumont, E., Vanhaecke, F., Cornelis, R., (2006). Selenium speciation from food source to metabolites : a critical review. *Anal. Bioanal. Chem.* 385, 1304-1323.

de Souza, M.P., Lytle, C.M., Mulholland, M.M., Otte, M.L., Terry, N., (2000). Selenium assimilation and volatilization from dimethylselenoniopropionate by Indian mustard. *Plant Physiol.* 122, 1281-1288.

El Mehdawi, A.F., Quinn, C.F., and Pilon-Smits, E.A.H., (2011). Effects of selenium hyperaccumulation on plant-plant interactions: evidence for elemental allelopathy ? *New Phytologist*, Vol. 191, pp:120-131.

Evans, J.C., Huddler, D.P., Hilger, M.T., Romanchuk, G., Mattews, R.G., and Ludwig, M.L., (2004). Structures of the N-terminal modules imply large domain motions during catalysis by methionine synthase. *Proc. Natl. Acad. Sci. USA*, 101: 3729-3736.

Eisenberg, D., et al., (1997) *Methods Enzymol.* 227, 396.

Ferrer, J.L., Ravenal, S., Robert, M., Dumas, R. (2004). Crystal structures of cobalamin-independent methionine synthease complexes with zinc, homocystenine, and methyl-tetrahydrofolate. *J. Biol. Chem.* 279, 44235-44238.

Fiser, A., Feig, M, Brooks, CL, Sali, A.(2002). Evolution of physics in comparative protein structure modeling. *Acc. Chem. Res.* 35, 413-421.

Finley, J.W., Sigrid-Keck A., Robbins R.J. and Hintze K.J., (2005). Selenium enrichment of broccoli: Interactions between selenium and secondary plant compounds. *Journal of Nutrition* 135, 1236-1238.

Finley, Y.W., Ip C., Lisk D.J., Davis C.D., Hintze K.J., Whanger P.D., (2001). Cancer-protective properties of high-selenium broccoli. *J. Agric. Food Chem.* 49, 2679-2683.

Gundampti, R.K., *et al.*, (2012). Protein protein docking on molecular models of Aspergillus niger RNase and human actin novel target for anticancer therapeutics. *J. Mol. Model.* 18, 653.

Guruprasad, K., Reddy, B.V., Pandit, M.W., (1990). Correlation between stability of a protein and its dipeptide composition: a model approach for predicting in vivo stability of a protein from its primary sequences. *Protein Eng.* 4(2), 155-161.

Hasanuzzaman, M., and Fujita, M., (2011). Selenium pretreatment upregulates the antioxidant defense and methylglyoxal detoxification system and confers enhanced tolerance to drought stress in rapeseed seedlings. *Biol. Trace. Elem. Res.* DOI:10.1007/s12011-011-8998-9, ISSN 1559-0720.

Ip, C., Thompson, H.J., Zhu, Z.J., Ganther, H.E., (2000). In vitro and in vivo studies of methylseleninic acid: evidence that a monomethylated selenium metabolite is critical for cancer chemoprevention. *Cancer Res.* 60, 2882-2886.

Iyer LM, Abhiman S, Aravind L (2011) Natural history of eukaryotic DNA methylation systems. *Prog. Mol. Biol. Transl. Sci.* 101: 25–104.

Koutmos, M., Pejchal, R., Bomer, T.M., Matthews, R.G., Smith, J.L., Ludwig, M.L, (2008) Metal active site elasticity linked to activation of homocysteine in methionine synthease. *Proc. Natl. Acad. Sci. USA*, 105, 3286-3291.

Kherraz, K., Kherraz, K., Kameli, A., (2012). Homology modeling of ferredoxin-nitrite reductase from *Arabidopsis thaliana. Bioinformation* 6(3), 115-119.

Keck, A.S., and Finley, J.W., (2004). Cruciferous vegetables: Cancer protective mechanisms of glucosinolate hydrolysis products and selenium. *Integrative Cancer Therapies*, 3, 5-12.

Lyi, S.M., Heller L.I., Rutzke M., Welch R.M., Kochian L.V. and Li, L. (2005). Molecular and biochemical characterization of the selenocysteine Se-methyltransferase gene and se-methylseleocysteine synthesis in broccoli. *Plant Physiol.* 138, 409-420.

Mount, D.W., (2001) Bioinformatics:sequence and genome analysis. *Cold Spring Harbor Laboratory Press,* Cold Spring Harbor, New York.

Marco Wiltgen, and Tilz, G.P., (2009). Homology modeling: a review about the method on hand of the diabetic antigen GAD 65 structure prediction. *Wien Med Wochenschr*, 159, 112-125.

Neha Arora, Vinay Kumar Singh, Kavitha Shah and Shashi Pandey-Rai, (2012). Qualitative and quantitative analysis of 3D predicted arachidonate 15-lipoxygenase-B (15-LOX-2) from Homo sapiens. *Biomedical Bioinformatics*, 8(12), 555-561.

Neuhierl, B., Thanbichler, M., Lottspeich, F., Bock, A., (1999). A family of S-methylmethionine-dependent thiol/selenol methyltransferase. Role in selenium tolerance and evolutionary relation. *J. Biol. Chem.* 274, 5407-5414.

Pedrero, Z., Madrid, Y., (2009). Novel approaches for selenium speciation in food stuffs and biological specimens: a review. *Anal. Chim. Acta*. 634, 135-152.

Pejchal, R., Ludwing, M.L., (2005). Cobalamin-independent methionine sysntehase (MetE): A face-to-face double barrel that evolved by gene duplication. *PLos Biol.* 3:e31.

Pilon-Smits, E.H.A., and Quinn, C.F., (2010). Selenium metabolism in plants. In: Cell Biology of Metal and Nutrients, Plant Cell Monnographs 17, R. Hell and R.R. Mendel, Springer Verlag, Berlin Heidelberg.

Ranocha, P., Bourgis, F., et al., (2000). Characterization and functional expression of cDNAs encoding methionine-sensitive and –insensitive homocysteine S-methyltransferases from Arabidopsis. *J. Biol. Chem.* 275, 1592-1568.

Rayman, M.P, Infante, H.G, Sargent, M., (2008). Food chain selenium and human health: spotlight on speciation. *Brit. J. Nutr.* 100, 238-253.

Soengas, P., Sotelo, T., Velasco, P., Cartea, M.E., (2011). Antioxidant properties of *Brassica* vegetables. *Funct. Plant Sci. Biotechnol.* 5, 43–55.

Sharma, P., Bhattacharya, A., (2012). 3Dstructure prediction of aquaporin-2, virtual screening and in silico docking studies of gold and silver derivatives, used as potent inhibitors. *Int. J.Pharam Pharm Sci.* 4(1), 574-579.

Suyama, M., et al., (1997).*J. Mol. Evol.* 1, S163.

Sali, A., Blundell, T.L., (1993). Comparative protein modeling by satisfaction of spatial restraints. *J. Mol. Biol.*, 234, 779-815.

Sanchez, R., Sali, A., (1997). Advances in comparative protein-structure modeling. *Curr.Opin. Struct. Biol.*, 7, 206-214.

Sors, T.G., Ellis D.R., and Salt D.E., (2005). Selenium uptake, translocation, assimilation and metabolic fate in plants. *Phytosyn. Res.* 86: 373-389.

Susan, J.F., Rachel, Collings, and Rachel Hurst, (2010). Selenium bioavailabity : current knowledge and future research reguirments. *Am. J. Clin. Nutr.* 91, 1848S.

Sors, T.G., Martin, C.P., Salt, D.E., (2009). Characterization of selenocysteine methyltransferases from Astragalus species *Plant J.* 59(1),110-122.

Thompson, J. D., T. J. Gibson, F. Plewniak, F. Jeanmougin and D. G. Higgins (1997) The CLUSTAL-X windows interface: flexible strategies for multiple sequence alignment aided by quality analysis tools. *Nucleic Acids Res.* 25, 4876-4882.

Traka, M., Gasper, A.V., Melchini A, Bacon JR, Needs PW, et al., (2008) Broccoli consumption interacts with *GSTM1* to perturb oncogenic signalling pathways in the prostate. *PLoS ONE* 3(7): e2568.

Verkerk R., Schreiner M., Krumbein A., *et al.,* (2008) Glucosinolates in *Brassica* vegetables: The influence of the food supply chain on intake, bioavailability and human health. Molecular Nutrition & Food Research. doi: 10.1002/mnfr.200800065.

Whanger, P.D., (2002). Selenocompounds in Plants and Animals and their Biological Significance. *J. Am. Coll. Nutr.* 21, 223-232.

White, P.J., Bowen, H.C., Marshall, B, Broadley, M.R., (2007). Extraordinarily high leaf selenium to sulfur ratios define "Se-accumulator" plants. *Ann. Bot.* 100, 111-118.

Willard, L., et al., (2003). VADAR: a web server for quantitative evaluation of protein structure quality. *Nucleic Acids Res.* 31,3316–3319.

Xiong, H., Buckwalter, B., Shieh, H., and Hecht, M., (1995). Periodicity of polar and nonpolar amino acids is the major determinant of secondary structure in self-assembling oligomeric peptides. *Proc. Natl. Acad. Sci. USA*, 92, 6349-6353.

Zhu, L., Jiang, C.J., Deng, W.W., Gao, X., Wang, R.J. and Wan, X.C., (2008). Cloning and expression of selenocysteine methyltransferase cDNA from *Camellia sinensis. Acta Physiol. Plant.* 30, 167-175.

INDEX

#

A

B

C

D

E

F

G

I

J

K

L

M

N

O

P

Q

R

T

U

V

W

X

Y

Z